The BMW Group home plant in Munich

BMW
GROUP

The BMW Group home plant in Munich

Caroline Schulenburg
Andreas Hemmerle
Susanne Tsitsinias

HIRMER

Contents

Foreword

Built to Shape Tomorrow

Munich is BMW's beating heart. Next to the Olympia Park, the home plant sits alongside the BMW Welt, the BMW Museum and the company headquarters – the so-called 'Four Cylinders' tower. The Munich site is completed by the nearby BMW Group Classic, the BMW Research and Innovation Center (FIZ) and the BMW dealership, as well as the exhibition pavilion in the city centre.

The plant's roots go back to 1913. Since 1922 the company has been manufacturing its products on these premises in the Milbertshofen neighbourhood – initially motorcycles and aero engines and then, from 1952, automobiles. The plant has developed in parallel with the city's urbanisation and is now in the heart of the city. The smooth interaction between the production technologies and supporting areas in a very confined space is a triumph in the automotive engineering world. The plant has always set an example for environmentally friendly and sustainable production – precisely because of its city-centre location.

Working with the FIZ, the BMW Group plant in Munich also serves as an internal specialist centre: from here, the process and technology expertise and the experience gained from 100 years of vehicle construction also extend to other production sites around the world.

This third version of our book is being published in the Munich plant's centenary year. It looks back on the plant's eventful 100-year history and portrays its exciting present. 'Built to shape tomorrow' is the centenary slogan. It underpins how capable of transformation the plant and its workforce are, their constant innovative strength and creativity. In recent years, for example, the company has prepared the home plant to the highest degree for the changes in the automotive industry by expanding its body shop and assembly, as well as constructing a resource-conserving paint shop. The combination of innovative and sustainable production structures and the transition to e-mobility with the BMW i4 has significantly strengthened the home plant's future viability.

The BMW Group's roots are fixed in Munich. History was and is made here. Our history lives in the buildings, production facilities and in the more than 7,000 employees from over 50 nations. Thanks to their hard work, commitment and insistence on perfection, BMW products are seen globally as something special. For the BMW Group home plant, the motto is: At home in Munich, at home in the world. With its unique architecture, history and extraordinary location, it is and will remain a Munich landmark, future forum and tourist attraction in equal measure.

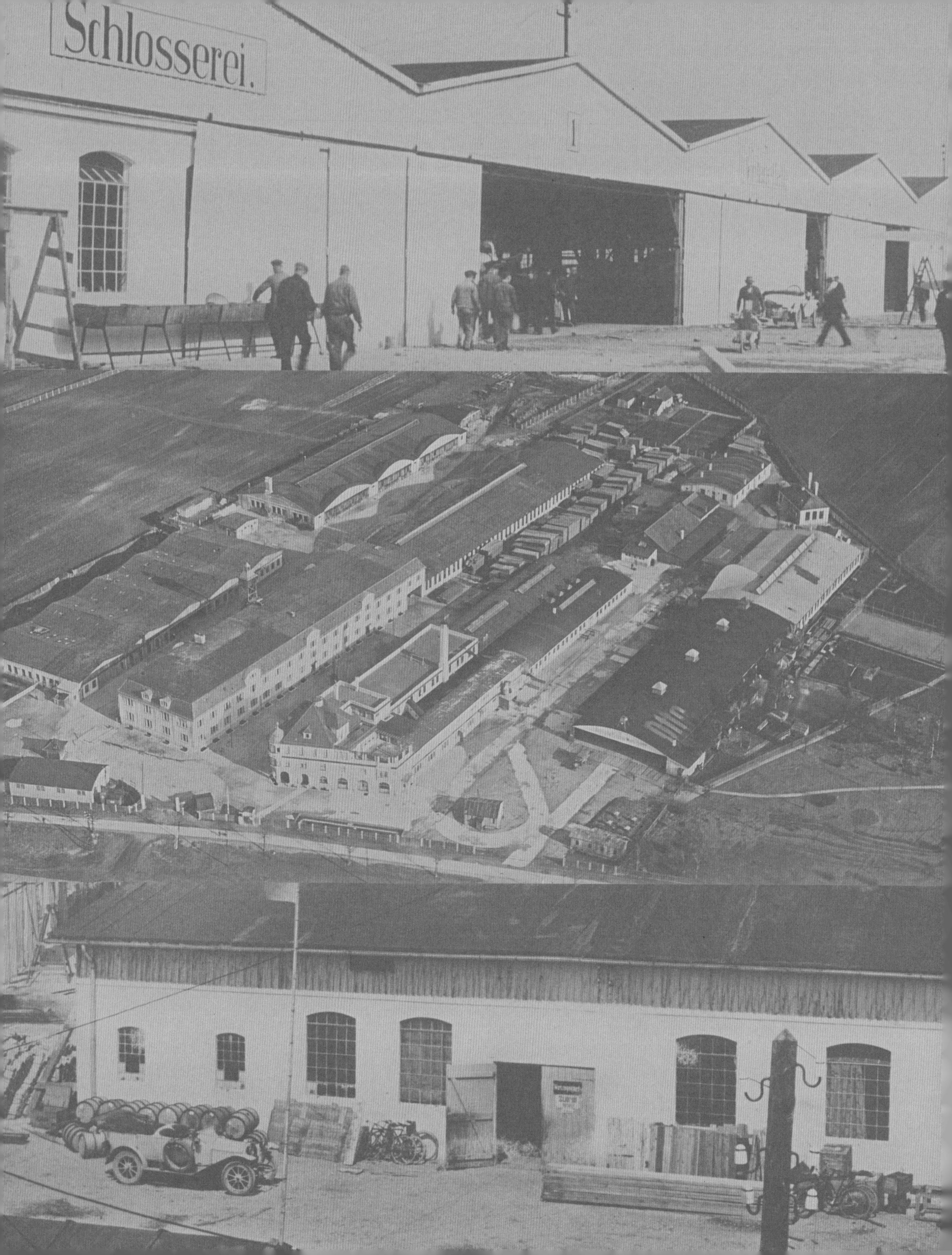
Schlosserei.

The origin of BMW's home plant

In 1913, aircraft manufacturer Gustav Otto began building a new plant at Munich's Oberwiesenfeld military base. In 1916, Bayerische Flugzeugwerke (BFW) took over the assets of the company, including the plant. The production halls were built using lightweight construction because of the constraints of wartime production, although this was only regarded as an interim solution.

1913 – 1920

Otto-Werke and Bayerische Flugzeugwerke (BFW)

The beginnings of BMW's home plant in Munich go back to the period before BMW actually came into existence. The land was originally acquired and developed by aircraft manufacturers "Flugmaschinenwerke Gustav Otto". Gustav Otto founded the company, which was named after him. He was the son of the inventor of the four-stroke internal combustion engine, Nikolaus Otto. Gustav Otto established his first production plant in Puchheim to the west of Munich before transferring operations to Schleißheimer Straße 135 in 1911. This move located the facility alongside Munich's Oberwiesenfeld military base. The base was at that point gradually being converted into an aerodrome. If a company wanted to do business with the military authorities, it was absolutely essential to establish a presence nearby. In addition to the facilities located in Schleißheimer Straße, Gustav Otto continued to use the production facilities in Puchheim, but he also

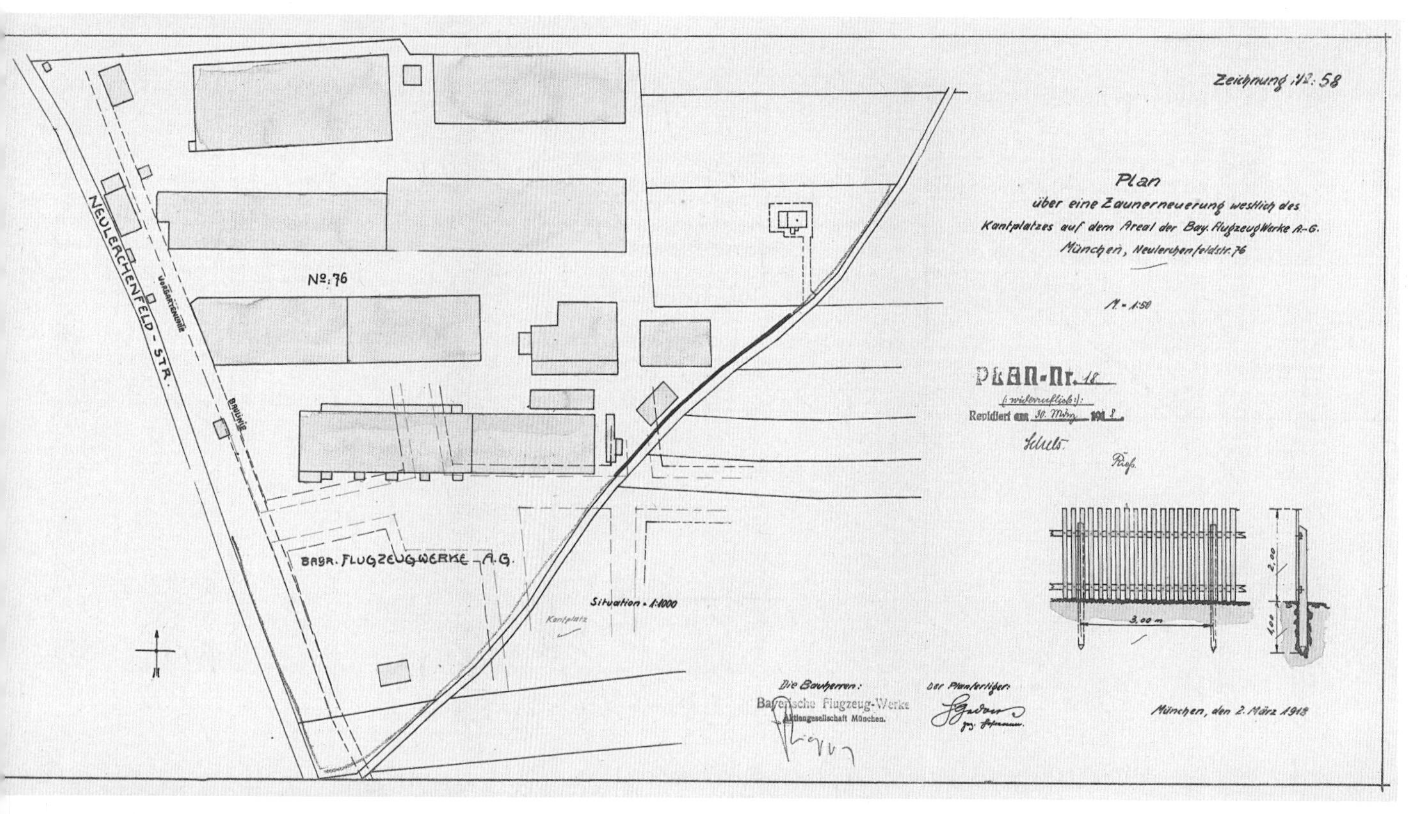

Site plan of Bayerische Flugzeugwerke AG, 1918.

Main gate at Otto-Werke on Neulerchenfeldstraße, 1915.

had two sheds in Neulerchenfeldstraße (later Neulerchenauer Straße) directly adjacent to the Oberwiesenfeld site.
At the beginning of 1913, Otto-Werke received an order for more than 40 military biplanes, and this meant that a new plant had to be constructed. Otto selected Neulerchenfeldstraße and started to build the plant the following year. The development was carried out by Munich construction company Ackermann & Cie. By June 1917, five large production buildings had been completed. They were configured from west to east and were built parallel to each other. The arrangement of the buildings allowed the aircraft to be rolled out of the hangars onto the airfield located to the west of the buildings without the need for any complex manoeuvres. The following buildings were constructed facing Neulerchenfeldstraße and ran from south to north: the pre-assembly building, the carpentry shop, the

warehouse and the assembly shop. A new aircraft hangar was constructed to the east of the warehouse.
The construction of the buildings was typical of the method used in the aircraft industry during the war. They were single-storey hangars built in lightweight construction, and the main material used was wood as this significantly cut down the construction time compared with methods using brick and steel. Since the plant was built during the First World War, the hangars had to be erected as quickly as possible in order to provide the company with the facilities to complete the orders placed by the military authorities. Another problem was caused by the difficulties of sourcing building materials, which also made it necessary to resort to wood. However, from the outset the plans envisaged that these buildings would only be used on a temporary basis. The aim was to replace the buildings with more robust structures after the war had come to an end.
Although Gustav Otto started the construction work at the plant, he was unable to make much use of the new premises. The financial situation of the company deteriorated steadily during the first two years of the war. This was partly due to the fact that the aircraft being produced at Otto-Werke had ceased to be competitive. Another factor was that, although Gustav Otto was a technically talented pioneer in the context of Bavarian aviation, he had little understanding of how to manage a factory on commercial principles geared to success in the marketplace. By the end of 1915, Otto-Werke was insolvent and the continuation of the company as a going concern was uncertain. However, since the military significance of aviation had increased enormously during the course of the war, the state was keen to have the maximum number of aircraft manufacturers. A government initiative therefore ensured that the shares of the company were incorporated within a new company, namely Bayerische Flugzeugwerke AG (BFW).
BFW and hence the plant located at Neulerchenfeldstraße expanded steadily. In 1917, a workforce of 2,400 people was already employed to manufacture up to 200 aircraft a month. However, the end of the war and the conditions of the Treaty of Versailles brought aircraft production to a standstill at BFW.

1913 Start of building work at the plant

1916 Bayerische Flugzeugwerke takes over the aircraft factories operated by Gustav Otto

1917 All production buildings are completed

Aerial view of Otto-Werke, 1915.

300

From BFW to BMW AG: the plant during the 1920s

In 1922, BMW AG transferred operations to the plant located in Neulerchenfeldstraße. During the course of the following year, the company started to manufacture a new product: the motorcycle. The increasing demand for aero engines and motorcycles necessitated expansion of the plant facilities from the mid-1920s.

1920 – 1930

Status in 1922 at the time of "transfer to BMW"

The history of the establishment of BMW AG proceeded in several stages. In 1917, Rapp Motorenwerke became BMW GmbH. Rapp Motorenwerke had been founded in 1913 by Karl Rapp. In order to expand the capital base of the company, it was converted into a joint stock company during the following year. After the end of the First World War, BMW faced a situation where there were no customers for the company's only product – the BMW IIIa aero engine – because all government orders had been cancelled. Production of aero engines was in any case prohibited under the terms of the Versailles Peace Treaty ratified in 1919/20. In June 1919, the company concluded a licence agreement with Knorr-Bremse AG to manufacture train brakes in order to keep the workforce in employment and ensure that the machines were not standing idle. In 1920, the sole owner of BMW AG, Camillo Castiglioni, sold his shares in the company to Knorr-Bremse AG. But two years later, he went back to Knorr-Bremse with an offer to purchase the engine construction facilities including all the technical

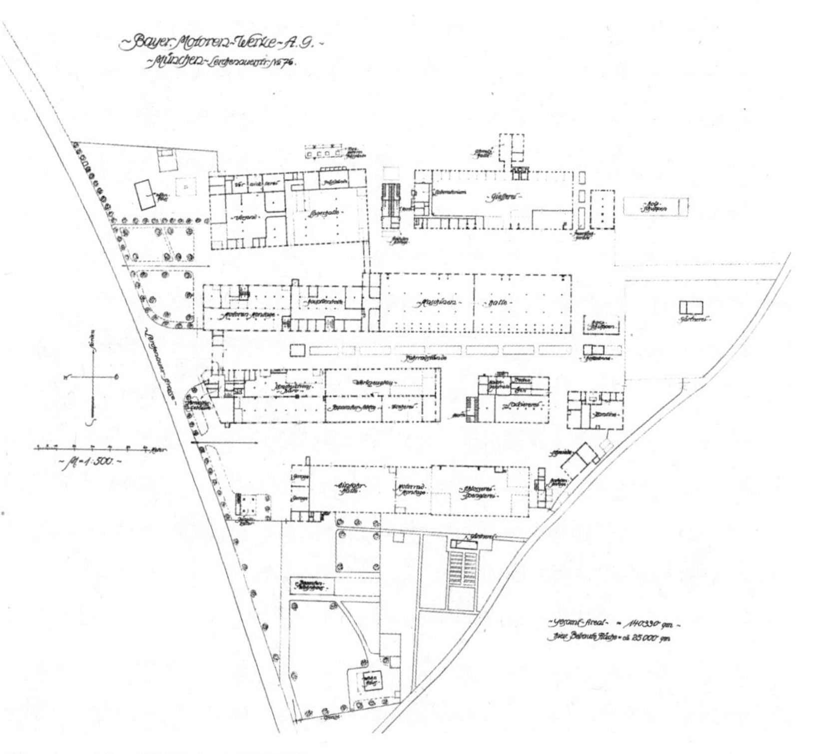

Site plan of the BMW plant, 1922/23.

Main entrance to BMW AG in Lerchenfeldstraße, 1922.

drawings, the machinery assets and workforce, and the company name BMW. At this point, Castiglioni was engaged in negotiations with the Czech Ministry of Defence for the possible manufacture under licence of the aero engines BMW IIIa and BMW IV. Licensed production of this nature offered the owner of BMW substantial opportunities for profit, and this was the reason for Castiglioni's renewed interest in the company. Since 1921, Bayerische Flugzeugwerke AG (BFW) with the factory in Lerchenauer Straße had been under Castiglioni's ownership, and he now transferred the shareholding he acquired to BFW. During the immediate post-war period, BFW manufactured motorcycles and wooden furniture, although the majority of the production buildings were empty. BMW engine production was therefore able to move into these facilities.

A triangular plot of land facing south was developed in 1922. This site was bounded by Lerchenauer Straße to the west and by a public footpath to the east. Six large production halls were available. The machine shop and the run-in facility/garage were located in the southernmost building. The building for assembly and inspection was sited to the north, together with a warehouse building. This in turn was connected with an engineering workshop to the east. The warehouse and dispatch department were built at the northerly end on the western side and the foundry was positioned to the east.

The year 1923 brought two changes at BMW AG: firstly, the Allies permitted limited reinstatement of aero engine production. Secondly, manufacture of stationary engines and installed units for different vehicles developed into a second mainstay for the fledgling company: engineers at BMW AG designed a motorcycle built around the "Bayern-Kleinmotor". The R 32 became the first vehicle to take to the road emblazoned with the blue and white BMW logo.

1917 Rapp Motorenwerke becomes BMW GmbH
1920 BMW loses its independence
1922 BMW AG is transferred to BFW
1923 Commencement of motorcycle production

Rebuilding of Hall 17

Starting up production of a new product – the motorcycle – and the return to manufacturing aero engines necessitated a number of changes at the factory site from 1923. The increase in orders also meant that the equity capital of BMW AG had to be increased several times. Some of the funds raised were used to expand the plant during the mid-1920s and, most importantly, to carry out a programme of modernisation.

In 1924, the company management decided to replace Hall 17, previously used for component production, with a new brick building. In order to avoid interrupting operations, the external structure in steel and bricks was erected around the existing wooden buildings. The new Hall 17 was built as a single space with a gallery running around the walls. The original wooden buildings located in the interior of the new structure meant that this gallery could not be supported by columns, so it was suspended from the roof structure instead. The wooden buildings were only demolished once the new production building had been completed. The machines were reconfigured to take account of the production workflow. The large Machine Hall 17 became the core of the BMW plant during the second half of the 1920s, and the following

Construction around Hall 17, 1924/25. The heart of the plant in the 1920s: Hall 17, 1926.

production facilities were located here: component production for aero engine manufacture and part of the motorcycle custom orders.
In 1928, a steel-framed building was constructed at the front and to the eastern end of Hall 17. The hardening shop was located on the ground floor and company offices were accommodated on the first floor. The hardening shop was an extremely important element in aero engine construction. In order to be able to withstand the high loads exerted on components during flight, certain components and materials had to be given special treatment. The hardening shop was equipped with oil-fired, gas-fired and electric furnaces of different sizes for standard heat treatments. It was also fitted with special equipment for carburisation and nitride hardening. Refitting the building allowed the most advanced technology to be used and created optimum working conditions. The interior of the building was 6.5 metres in height and this was combined with a powerful ventilation system to ensure that heat and hazardous vapours were extracted from the building as quickly as possible.

1924 Commencement of expansion at the plant

1924 Construction around Hall 17

1928 Construction of a new hardening shop

The foundry

The increasing volume of production, especially in motorcycles, necessitated a number of changes and an enlargement of the factory. In 1928, a new foundry was built. Already in 1918 high-quality products were being manufactured at the BMW foundry, and this reputation was the reason why other companies were sourcing cast components from BMW. Construction of the new building took account of all the latest statutory heath and safety requirements. The electron-beam foundry was located separately from the other departments in order to protect associates from the sulphur dioxide emissions produced. The new foundry was equipped with the most advanced production facilities available at the time. It included departments for aluminium, electron-beam, iron and bronze casting and the associated smelting chambers, as well as a facility for cleaning up all the castings. A model-makers' workshop, offices and recreation rooms were arranged around a gallery in the hall. Sand treatment machines were positioned at the western and eastern ends of the foundry, with two conveyor belts taking the sand to the bunkers that were distributed along the entire length of the building. From here the sand was then delivered directly to the moulding machines. The smelting building for producing grey-cast iron was connected to the eastern end of the sand treatment room located to the west. The building with the cleaning facility for the castings was positioned alongside. The aluminium smelting building was also located facing the east. Access to the electron-beam foundry from the aluminium smelting building was through the sand preparation room for the aluminium foundry. The electron-beam foundry extended over the entire height of the building and was completely separated from the other rooms in the foundry. The core-making facility was located at the northern end of the building and was supplied with sand in the same way as the moulding machine.

The new foundry, 1929/30.

Sand bunkers in the foundry, c.1930.

Inside the new foundry, 1930.

The test track

In 1927, a new motorcycle test track was constructed on the northern periphery of the site along what was then Keferloher Straße. Up until that point, the quality of the motorcycles rolling off the production line had been tested inside a building. A circuit 750 metres in length was now available on which the test riders were able to take the machines to a top speed of 140 km/h. New aero engine test rigs were located inside the circuit. They were sunk into the ground to a depth of around four metres. A tunnel passed under the test circuit to link the test rigs with a new two-storey building where aero engine reassembly was located on the ground floor. This configuration created short routes so that the quality of the tested engines could be checked in the reassembly area as quickly as possible. The apprentices' workshop was initially located on the first floor of the building. The testing department was sited to the west of the reassembly facility and access from the test circuit was also through a tunnel. The new logistics

Running in BMW motorcycles, 1929.

1927 Construction of a new test track

Aerial view of BMW's Munich plant with the test track in the background, 1928.

saved time because the motorcycles could be washed, checked and prepared for shipment immediately after they had been tested. The dispatch department was also located in this building. A rail spur to the complex was added at the same time as the test track, allowing the checked and crated motorcycles and engines to be shipped directly by the dispatch department.

Aero engine production

The most important product manufactured at BMW up to 1945 was aircraft engines. Because aero engine production initially had to be suspended after the end of the First World War, BMW engineers set about turning the existing registered models of aero engines into power units for other purposes. The company hoped that the good reputation of the first BMW unit, the BMW IIIa, would ensure that the newly developed marine, lorry, automobile and motorcycle engines would generate good sales. Sadly, these expectations proved to be unfounded and the engines brought in virtually no profit for the company. Partial revision of the Versailles Treaty in 1922 once again opened the door for aero engine production. However, the main customers for BMW engines were not in the German market: major orders came mainly from the Soviet Union, the Czech Republic and Japan.
Aviation during the period between the wars focused on ongoing development of aircraft which required more powerful engines and consistent operational safety. Instead of pursuing costly new developments, engineers at BMW concentrated on developing and refining the proven models. In 1928, BMW also acquired a licence from American aircraft manufacturer Pratt & Whitney to manufacture an air-cooled radial engine.
Right from the start, BMW aero engines were built to extremely high quality standards. In order to maintain this level of quality, BMW always took great care to ensure that its associates were highly trained and appropriately

Designers at the drawing board, 1925.
Left: BMW VI aero engines in the reassembly facility, 1926.

qualified. Power units were also subjected to a system of continual quality checks and inspections. Already during the production process, most of the materials used were subject to very precise testing. The important role of quality in the production process is reflected in the incoming goods inspection that checked raw materials and all outsourced parts and components. But such inspection was not simply carried out in the preparatory stages prior to production: checks were also implemented during the actual manufacturing process. When associates were being trained in a new area of activity, their first products were put through a rigorous testing process in a "first article inspection", although workers often regarded this as unnecessary "nannying". However, the inspection established that all the instructions issued by the master craftsman had been understood and were being properly implemented. Such checks reduced the level of rejects and enhanced the quality of the engines.
Once the quality of the individual components had been subjected to rigorous testing, a great deal of effort was also put into ensuring that the engines lived up to the high standards. This is why only specialist craftsmen were normally used to assemble engines. Since monthly production figures during the 1920s were never higher than around 50 units, volume production was not an issue. The power units were therefore assembled as specials or in groups of two or three. Once assembly of the aero engines had been completed, the units were put through the final inspection, which constituted the last stage of production. Initially, each engine was run for several hours on a test rig. It was then returned to reassembly and completely taken apart. After all the components had been cleaned in a petrol bath, they were examined for material faults or damage, such as cracks. After ascertaining that all the components were in perfect condition, the engine was reassembled. It was then run at maximum power on the test rig for another two hours. Only when the engine had been taken apart again to establish that all the components were perfect and free of faults was it put back together and crated ready for shipment. Quality specifications to these exacting standards were normal in the aero industry, and they were absolutely essential because engine defects during operation could have untold consequences.

1920 Prohibition on aero engine production by the Versailles Treaty

1922 Partial amendment of the Versailles Treaty

1928 Acquisition of the licence to manufacture an air-cooled radial engine from Pratt & Whitney

Advertisement highlighting the different uses for BMW engines, 1920.

Dornier "Wal" flown by Wolfgang von Gronau after landing in New York, 1930.

The BMW VI aero engine

During the first half of the 1920s, BMW manufactured six-cylinder engines – the BMW IIIa and BMW IV units – before going on to manufacture a 12-cylinder engine, the BMW VI. This model allowed BMW to meet the requirements of the market for more powerful engines. The BMW VI was a design derived from the BMW IV. The principle of developing existing products was typical for BMW at that time and the two engines contained many identical parts. During the summer of 1924, the first BMW VI engines were run on the test rigs. Two years later, the first power unit was put through its first type inspection at the German Laboratory for Aviation (DVL). When production came to an end in 1937, BMW had produced around 6,000 units of the engine. Numerous record and long-distance flights were achieved with the BMW VI. In 1930, for example, Wolfgang Gronau undertook the first crossing of the Atlantic from east to west in a flying boat, in a Dornier "Wal" powered by two BMW VI engines. Numerous variations in the design of the BMW VI engine were produced and the engine was also manufactured under licence in Japan and the Soviet Union..

In 1924 the R 32 was still built without flow production techniques.

Motorcycle production

After the individual components for the motorcycles had been manufactured, they were first inspected in the parts warehouse and subsequently transferred to the assembly departments. On the assembly lines, gearboxes, front forks and rear-wheel drives were assembled from individual components to form assemblies. After testing these were then ready for mounting in motorcycles. Each assembly line came to an end precisely at the point on the final assembly line where the incoming assembly had to be incorporated into the frame of the motorcycle, which was mounted on an assembly dolly. The engines were in turn assembled on an assembly line where they were mounted on dollies and passed on from one worker to the next before being transferred to the engine testing area. Finally, each power unit was put through a test run on the test rig. The fully assembled motorcycles were then transported to the running-in department where they were tested for smooth-running operation. This process was often carried out by passionate and experienced motorcycle riders who could "sense" even the smallest defect. Construction of the test track around the end of the 1920s allowed the motorcycles to be tested over a longer distance and at higher speeds.
The BMW R 32 marked the start of a long series of high-quality BMW motorcycles. Up to the beginning of the 1930s, motorcycles and aero engines were grouped in a single production unit at BMW. This was necessary

due to the very erratic flow of orders for aircraft engines. In order to provide continuity in the supply of work for the highly qualified specialist fitters necessary for aero engine production, the workers were employed in motorcycle manufacture when aero engine production was slack. Organising the manufacturing process in this way meant that the production costs of BMW motorcycles were so high that they numbered among the most expensive motorcycles in the world. By the same token, the standard of quality was well above the average for the market, which meant that the company was able to generate profits from motorcycle production.

The range of motorcycles produced at BMW AG was continuously expanded during the 1920s. Production of the R 37 sports model, developed from the R 32, started up in 1924. The single-cylinder R 39 motorcycle came on stream during the following year to complement the production programme at the entry-level end of the range. However, the high price of this model proved to be uncompetitive in this market segment and production was phased out as early as 1926. By contrast, demand for the flat-twin "boxer" models was extremely high. This encouraged BMW to steadily expand the product range during the 1920s. The product portfolio always included touring and sports motorcycles. This differentiation was based on the principle of "same frame – modified engine".

Promoting the quality and reliability of BMW motorcycles, 1929.

R 52 solo motorcycles and sidecar combinations in the Reichswehr army version ready for shipment, 1928/29.

Above: The start of the assembly line where the frame and fairings were assembled for the BMW R 52 and BMW R 62 models. Completed engines can be seen to the right of the line, 1928.
Below: BMW R 32 motorcycles in the dispatch building, 1924.

Just like BMW aero engines, the motorcycles manufactured by the Munich company stood for high quality, advanced technology, dependability and sportiness. Since BMW also succeeded in bringing down sale prices through rationalisation measures, the demand for BMW motorcycles in Germany and abroad rose steadily during the 1920s. During the first four years from 1923 to 1927, production increased from 1,500 to around 5,000 units.

The growing demand for motorcycles meant that the manufacturing facilities available at the plant for this product had to be expanded. When the major programme of expansion was implemented at the end of the 1920s, a new building for motorcycle production was constructed to the south of the foundry. Anyone entering the building from the western side first came to the department for frame construction. The frame components for the new models made of pressed steel sheet were manufactured at Fahrzeugfabrik Eisenach, the automotive plant which BMW acquired in 1928. They were simply riveted together in Munich before being sand-blasted in separate rooms, painted and then transferred to the assembly line. The main part of the building was used for the production of individual components. Special machine tools for motorcycle construction were set up there.

1923 Launch of the R 32

1924 Launch of the R 37

1927 Annual motorcycle production reaches 5,000

The BMW R 32

In 1923, BMW launched a new product range with the R 32. The first BMW motorcycle to be manufactured already had a boxer engine mounted transversely to the direction of motion and a drive shaft for transmission of power to the rear wheel. These two features are characteristic of BMW motorcycles to this day. The quality of the R 32 was impressive: all components that were liable to require repair were encapsulated, and the drive shaft required less maintenance than the chains or belts that were otherwise common at the time. The engine was designed to generate a very low output of only 8.5 hp and lived up to BMW's reputation for reliability. The chassis with the double-tube frame and the short swinging arm at the front was better suited to the theoretical top speed of 95 km/h than the condition of the roads at the time. With the streamlined and sturdy R 32, BMW had laid the foundation stone for an impressive future of motorcycle development.

Group photo with BMW M 2 B 33, 1925.

The BMW workforce

The high level of qualification attained by BMW associates meant that the workforce was regarded as a vital part of the company right from the start. The significance of the specialist workforce is also evident from the fact that Camillo Castiglioni not only acquired the company name, the trademark, patents and design drawings from Knorr-Bremse, but most importantly also the specialist workforce. The high standards entailed by aero engine production demanded a significant proportion of highly skilled staff in the workforce, including engineers, master craftsmen, fitters and specialist workers. Despite the two areas of activity encompassing motorcycle and aero engine production, the company was not always in a position to keep the entire workforce busy throughout the year. The winter months in particular saw seasonal lay-offs in motorcycle production. Employment in aircraft engine production was highly dependent on big orders. The period of notice for workers at that time was around four weeks, so that the company experienced no problems under employment law if they dismissed staff due to low order books. Since the aero engine mechanics were generally regarded as the "kings among mechanics", BMW made strenuous efforts to retain as many highly qualified specialist workers as possible even if order books were low. However, the demand for motorcycles and aero engines fell during the global economic crisis and this inevitably impacted on the workforce. For example, standard working time was reduced by one sixth from 48 to 40 hours in 1931. When the economic situation in Germany deteriorated still further, the government issued an emergency decree in October 1932, cutting wages by up to one fifth. This led the workforce in Munich to take strike action. The company management reacted by dismissing all the strikers, but industrial action was brought to an end by a compromise after a few days.

BMW workers standing on a petrol tank, 1927.

557

Plant expansion to meet increased demand for aero engines in the 1930s

The BMW plant underwent further expansion throughout the 1930s. From 1933 onwards, BMW increased production capacities for aero engines in order to meet growing demand. In 1934, the company's entire production of aero engines was transferred to BMW Flugmotoren GmbH. Between 1931 and 1939 the workforce also experienced a marked rise in the number of associates and, in order to attract workers at a time of full employment, BMW AG broadened its provision of occupational welfare.

Physical expansion of the plant

By the early 1930s the plant was already undergoing enormous expansion. Planning for this went as far back as 1926, a fact documented by pre-construction drawings from that year. By buying up plots of land to the south-east of the premises the company was able to build a U-shaped complex of buildings on that side of the plant. These consisted of a machine shop and assembly hall for aero engine construction, as well as a warehouse building. Enlargement of the plant was necessary in order to meet incoming bulk orders for aero engines destined for the development of a German air force. By the mid-1930s all the land between Keferloher Straße in the north, Riesenfeldstraße in the east, Neue Straße (later Dostlerstraße) and Lerchenauer Straße in the west had been used for construction purposes. This expansion also brought about a change to the plant's axis. Where formerly it had run east to west, now the new main entrance located at the southern end and construction of a new administration building meant alignment of the plant was shifted to a north-south axis. As plant developments in recent years had been concentrated mainly on the eastern side of the premises, the new north-south axis now ran right through the centre of the plant. The construction of a new main entrance

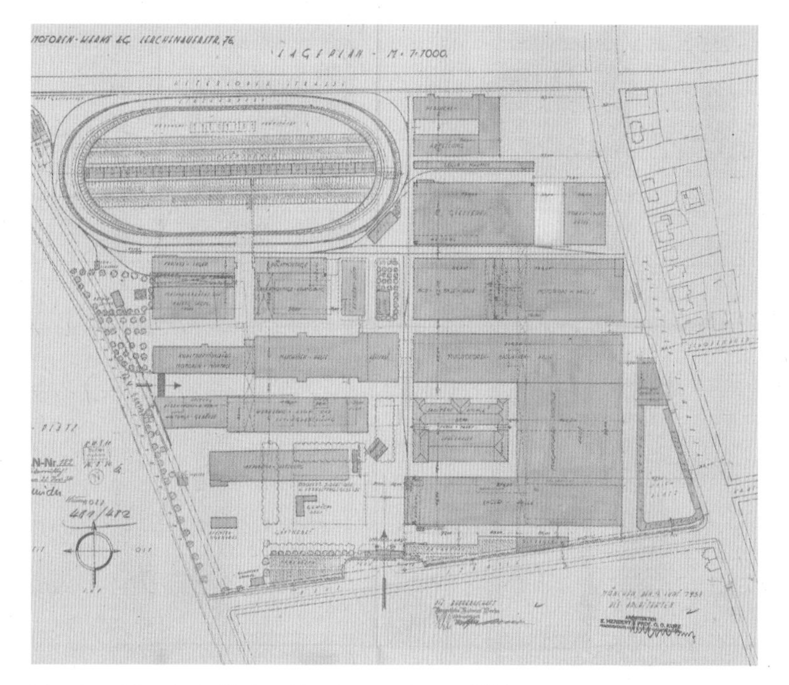

Site plan of the plant with the old and new entrance, 1934.

was inevitable since Lerchenauer Straße had now become too narrow for delivery vehicles and could no longer meet contemporary requirements. The growth of orders in aero engine production also called for an increase in capacity. So in 1934 BMW AG decided to set up BMW Flugmotoren GmbH. This company covered all BMW's activities in the field of aero engine production. It was also a development that enjoyed the full backing of the Reich government. In Germany, GmbHs (limited liability companies) were required to publish much less information than AGs (stock corporations), so creation of the new company would help provide cover (at least initially) for Germany's secret armaments programme. In order to enlarge production facilities, BMW acquired an additional plot of land to the north of Keferloher Straße in the 1930s and ceded it to BMW Flugmotoren GmbH in the form of an inheritable building right. The plot returned to BMW AG ownership in 1947. The second half of the 1930s also saw several new buildings erected on the northern site. As a result of land acquisitions and plant expansion throughout the 1920s and 30s, the BMW home plant grew to cover roughly the area it occupies today.

1934 Foundation of BMW Flugmotoren GmbH

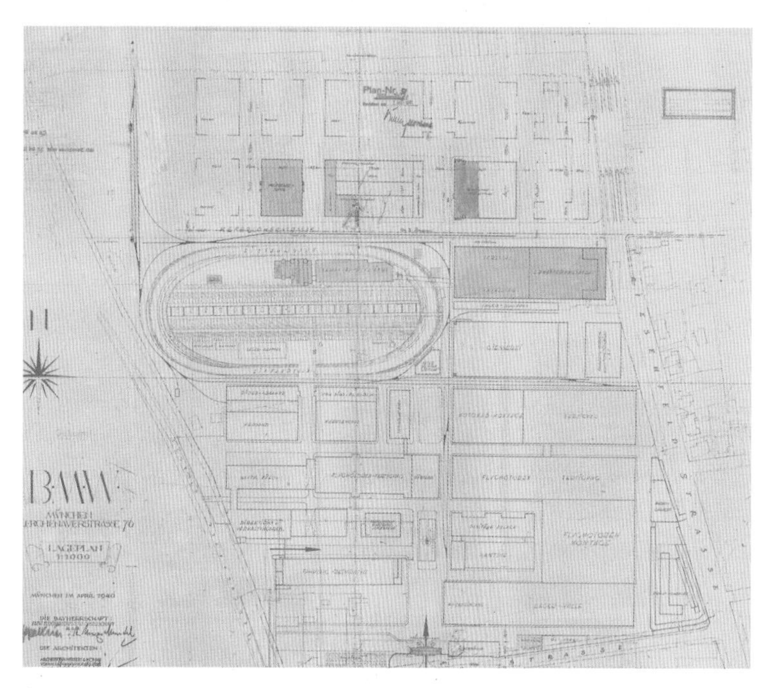

Site plan with the acquired land to the north, 1940.

Final assembly of the series-produced BMW 132 aero engine, c. 1940.

Aero engine production

The armament policies of the National Socialist regime required an ever-increasing supply of aero engines from 1933 onwards. To satisfy this demand, the Munich plant built two different aero engine models during the 1930s: the tried-and-tested water-cooled BMW VI, production of which continued as before; and from 1934 the air-cooled BMW 132 radial engine. In addition to the development of new plant locations, the company also adapted plant production processes to meet growth in demand. Instead of the single or group production approach common in the 1920s, for aero engines BMW switched to series production. But with the growing complexity of aero engine technology this also meant simplifying the production process to a series of individual steps, while at the same time observing strict quality control standards. From the mid-1930s, however, the company was increasingly forced to recruit unskilled workers. So in order to ensure that such specialist work was carried out without error, BMW was quick to develop a series of comprehensive training and upgrading programmes. Before being given a job in production, new workers were required to pass a 12-week training course.

1934 Start of series production of the BMW 132

Production of the air-cooled BMW 132 radial engine, c.1934–37.

A BMW 132 on the test rig, 1932.

The BMW 132 aero engine

After acquiring a licence to produce air-cooled radial engines from the American Pratt & Whitney Aircraft Company on 3 January 1928, BMW began by manufacturing the nine-cylinder Pratt & Whitney Hornet model as the "BMW Hornet". But the relevant authorities in Germany preferred water-cooled engines. As a result, BMW sold relatively few Hornet engines and the licence agreement was terminated by mutual agreement in 1931. It was only when the authorities demanded that Junkers JU 52 aircraft should be fitted exclusively with BMW radial engines that demand began to grow. BMW then renewed its licence agreement with Pratt & Whitney. In 1933 the company received design plans for the Hornet S4D2 model and marketed this design under the designation BMW 132. Various versions of this engine were built, individual models being identified by a letter added to the type designation. The company built the 132 A, D, E, H and L carburettor versions, as well as direct injection versions with the designations 132 F, J, K, M and N. Depending on the specific version, these engines delivered up to 1,000 hp. The use of the BMW 132 in the Junkers JU 52 – still a popular aircraft today – was a key factor in the engine's high production volume throughout the 1930s.

Motorcycle production

The purchase of Fahrzeugfabrik Eisenach, the vehicle factory in Eisenach, in 1928 enabled BMW to begin continuous flow production. This "American style" of production had hitherto been unnecessary for aero engines and motorcycles on account of the low unit numbers required. In the 1920s, maximum annual motorcycle production was around 5,600 units. But the end of the global economic crisis from mid-1932 onwards brought increased demand for motorcycles. More rational production methods gave productivity a considerable boost. When 10,000 units were produced in 1935, the five-figure production threshold was crossed for the first time. And in the year before the outbreak of war, annual production of motorcycles rose to 17,017. In order to improve profit margins in the motorcycle division it was decided in the late 1920s to pursue a new model policy. Until then BMW had brought out a new model programme almost annually. For the 1930s, however, it was decided that although motorcycle models would be subjected to regular revision, the model designations would apply over a longer period. The new models used pressed steel, which not only gave the frames greater durability, but also brought about a tangible reduction in manufacturing costs. The combination of pressed steel frames and boxer engines was soon adopted by other German motorcycle manufacturers – which explains why contemporary observers were soon talking of a "German school" of motorcycle construction.

1935 Annual motorcycle production rises to around 10,000 units

BMW R 12 motorcycles, destined for export to China, lined up in the dispatch building, 1937.

The BMW R 5

At the Berlin Motor Show in spring 1936, BMW presented an all-new motorcycle in the shape of the R 5. The frame was designed from electrically welded conic oval tubing. This light frame in combination with the revolutionary telescopic fork promised outstanding handling qualities, even if the rear wheel was still unsprung. The engine, too, was a completely new design. Double camshafts driven by a timing chain took care of valve operation. Fuel was prepared in two carburettors, each fitted with a small air filter. However, these "ear" filters were highly susceptible to dirt and were replaced in 1937 by a central air filter on the transmission casing. The four-speed transmission was now operated using the left foot, although there was also an auxiliary hand gearshift lever on the right-hand side of the transmission block to help less experienced riders select gears more easily. A further innovation was the foot brake, which was no longer activated with the heel but – as is common today – with the ball of the foot.

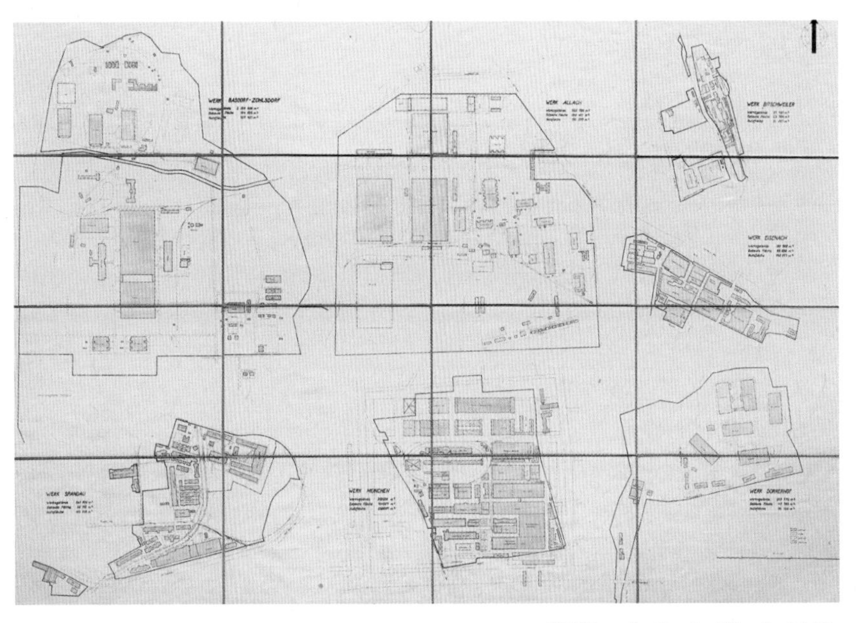

BMW production facilities in 1942.

Other BMW plants

During the 1920s and 30s, BMW acquired new production facilities in addition to the home plant in Munich-Milbertshofen. In 1928 the company bought the Fahrzeugfabrik Eisenach and made its first foray into car production. Initially this plant continued to produce Austin Seven vehicles under licence as it had done before the BMW takeover. In 1932 the company unveiled its first car developed in-house, the BMW 3/20 PS. Then throughout the 1930s, as part of the general armaments policy, other plants were built for the purpose of aero engine production. In 1936 BMW established a plant in Allach in the north of Munich and another in Dürrerhof near Eisenach. Both production facilities were known as "forest factories" or "shadow factories". On the instructions of the Reich government, the company built these plants outside the cities in order to camouflage their location. In 1939 BMW acquired Brandenburgische Motorenwerke (Bramo) in Berlin-Spandau from Siemens-Halske. Facilities were also built in Basdorf and Zühlsdorf as supply plants for Spandau.

1928 Acquisition of Fahrzeugfabrik Eisenach

1936 Rebuilding of the Allach and the Dürrerhof plant

1939 Purchase of Brandenburgische Motorenwerke (Bramo)

Occupational welfare provision

BMW made a name for itself from very early on with its occupational welfare services for associates. From the outset the plant premises boasted a canteen, which was considerably enlarged in the early 1930s. Similarly, sanitation facilities such as the washrooms underwent renovation. BMW also constructed housing for salaried associates on the plant premises and apartments for workers close by.

At the outbreak of the Second World War, BMW in Munich was one of the first German companies to set up its own company health scheme. An in-house works doctor was appointed, whose responsibility was to maintain the health of associates (by undertaking, for example, precautionary health checks) and to treat work-related accidents on site. From the mid-1930s BMW used occupational welfare as a means of attracting new associates in a climate in which labour was becoming increasingly scarce. Furthermore, BMW was unable to recruit new workers by raising wages since the government had also regulated wage policy.

One of the last occupational welfare services to be established at the BMW plant in 1940 was the day nursery. This measure was taken largely for reasons of employment policy. The size of the available labour force shrank dramatically at the outbreak of the Second World War, when most male workers were drafted into military service. In view of this, companies turned to women to fill jobs, even though they lacked the specialist skills of trained workers. But since National Socialist thinking saw women predominantly in the role of wives and mothers, it was also necessary to reconcile ideology with a pragmatic solution. Thanks to child-care nurseries for babies and young children, even young mothers could work on the production lines without problem. Attention was also paid to contemporary medical advice which asserted that the mortality rate among babies breastfed for only a short time was markedly higher than that among infants weaned on their mother's milk over a longer period. Mothers were therefore permitted to breastfeed their children during working hours.

1939 Establishment of a health scheme

1940 Opening of the BMW day nursery

Above: Laboratory at the sanatorium, 1941.
Middle: The dietary kitchen in the new canteen, c.1935.
Below: Garden of the day nursery, 1942.

The day nursery was established as a separate building with garden directly adjacent to the plant. The nursery took in babies and young children up to the age of three. Contemporary accounts depict a haven of tranquillity – "a building flooded with daylight", "friendly dormitories" and "comfortable breastfeeding facilities". Only references to the obligatory air-raid shelter gave any indication of the wartime context.

Washroom of the day nursery, 1942.

Outdoor work break in the 1930s.

BMW associates

The armaments policy of the National Socialists brought about a steady rise in the production of aero engines – a development that also had an impact on BMW AG and the plant at Milbertshofen. With the enlargement of the premises at Flugmotoren GmbH as a result of the acquisitions to the north already documented, the size of the workforce grew markedly throughout the 1930s. Employee numbers rose from around 1,300 in 1931 to 9,745 in 1939. Out of a total of 5,884 workers and salaried associates in 1935, the aero engine division had 4,416 associates, with the remaining 1,468 occupied either in motorcycle production or in administration. Production of aero engines also increased accordingly from 257 units in 1931 to 1,986 engines in 1939.

When the National Socialists took power, associate representation at BMW AG was brought into line accordingly. This meant, for example, that works council representatives – often with many years of experience in the role – were forced to step down in favour of NSDAP party members, appointed to the rank of "works steward" in accordance with recently imposed social legislation. The role of the trade unions was assumed by "subdivisions" of the NSDAP. In 1934 the German Labour Front (DAF) called for the abolition of clocking on and off using a time clock and tried to replace it instead with a roll call held at the beginning and end of each shift. But BMW resisted on the grounds that this would result in the loss of valuable working time and disturbance to engine production. For this reason roll calls were conducted only on special occasions.

Roll call for the Munich plant at the launch of the Winter Aid Charity (WHW), 1936.

300

The plant during the Second World War

In the course of the war BMW developed into a dedicated armaments manufacturer. When motorcycle production was transferred to Eisenach, the home plant in Munich was devoted solely to the manufacture of aircraft engines. BMW began using forced labour in production in 1940. In 1943, Allied bombing raids compelled the company to move its production facilities elsewhere.

1939 – 1945

War-related changes

At the outbreak of the Second World War, the production of vehicles for civilian use was greatly curtailed. As a result, BMW developed into a dedicated armaments manufacturer and the Milbertshofen plant increasingly served the purposes of the military. Within the network of BMW plants, the Munich plant was responsible for developing and manufacturing the numerous variants of the BMW 801 engine. The Milbertshofen plant was also home to the BMW 802 and BMW 804 projects. These designations stood for 14- and 18-cylinder engines for combat and fighter aircraft. From 1940, the almost exclusive dedication to aircraft engine production and the increasingly large orders placed by the Reich Ministry of Aviation resulted in a redefinition of tasks for all BMW plants. As part of this reorganisation, the Munich site was designated a "pilot production plant". The idea behind the pilot production plant was to carry out test production runs and determine whether the models built during the development phase could be implemented in large-scale production without any hitches. Faced with an acute scarcity of skilled workers, the company had to introduce work processes and steps that were as short and simple as possible. These were then tested in the pilot production plant. The grounds of Flugmotoren GmbH in the northern section of the Munich plant became the home of the

Front view of the BMW 801 C twin-row radial engine, 1942.

development departments of BMW Flugmotorenbau GmbH Munich and BMW Flugmotorenwerke Brandenburg GmbH, which were merged in 1942 at the behest of the Reich Ministry of Aviation. Only the development group for jet and rocket engines remained in Spandau and Zühlsdorf in the northeast of Berlin. Geographic concentration reflected the focus of activities: all development work at the Munich plant was directed towards the refinement of the BMW 801 engine.

However, limited construction capacities meant that the expansion of the pilot production plant was postponed repeatedly during the war and never actually came to fruition. Furthermore, trial production orders had to be postponed until 1943 because expansion of the high-volume production plant in Allach was delayed and the Munich facilities had to be used for large-scale production. In addition, the local planning office demanded that the expansion of the plant fit within the overall framework of Munich as a strategic hub.

In order to obtain further production capacities for the manufacture of aero engines, BMW moved motorcycle production from Munich to Eisenach at the request of the Reich Ministry of Aviation in 1942. As a result, the company gained some 20,000 square metres of floor space that could be used for the production of aircraft engines without having to undertake any building work. At the same time, relocation of motorcycle production meant that the Munich site could concentrate solely on aero engine production, though this was viewed with some misgivings within the company. As a result of the relocation, production facilities were no longer available at the Munich site for non-military production.

The BMW 801 aero engine

The BMW 801 was the only German twin-row radial engine to be produced during the Second World War. The first air-cooled aircraft engine developed exclusively by BMW, it had 14 cylinders and an output of up to 2,000 hp. The first test runs in 1939 were supposed to be followed by extensive test series. However, after the Nazi regime's invasion of Poland sparked off the Second World War, the Reich Ministry of Aviation ordered the rapid delivery of BMW 801 engines. Large-scale production was hastily started, throwing up various technical problems that ultimately could not be resolved until 1942. By the end of the war the BMW 801 was one of the most widely used German aero engines around, having been installed in countless variants in a broad range of aircraft. The BMW 801 is particularly notable in that it was the first aero engine whose exhaust gas turbocharger reached mass production maturity. Another special feature of the engine was a master control unit that enabled the pilot to control all the engine settings with a single lever.

Relocation of production

From March 1943, bombing of the Munich factory became more intense. The company's management had long intended to build bombproof bunker systems but the plans had not materialised. Starting in late 1943 it was therefore necessary to move production to safeguard its continuation. The initiative to relocate production came primarily from the Reich Ministry of Aviation. After several inspections of possible new sites with ministry representatives, it was decided that spinning and weaving facilities offered the best technical conditions. All dispersal operations and sites were assigned code names. However, since the production machinery was transported in trucks with "BMW Munich" emblazoned on the sides, it is questionable whether the operation actually remained secret.

The relocations were accomplished in three partly overlapping phases. Plans were finalised at the end of 1943. BMW carried out the interim relocation operations in the spring and summer of 1944. In 1944 production facilities were moved back to Munich. This was followed almost immediately by a new, highly chaotic relocation. The production moves exacted a high price in terms of time and costs. Because the workforce had to be moved along with the machinery, accommodation had to be arranged for the workers. The outlying production was centrally organised and controlled. Many of

The relocation of machinery could hardly be disguised, 1944.

1940 Reorganisation of the BMW Group plants

1942 Relocation of motorcycle production to Eisenach

1942 Concentration of development departments in Milbertshofen

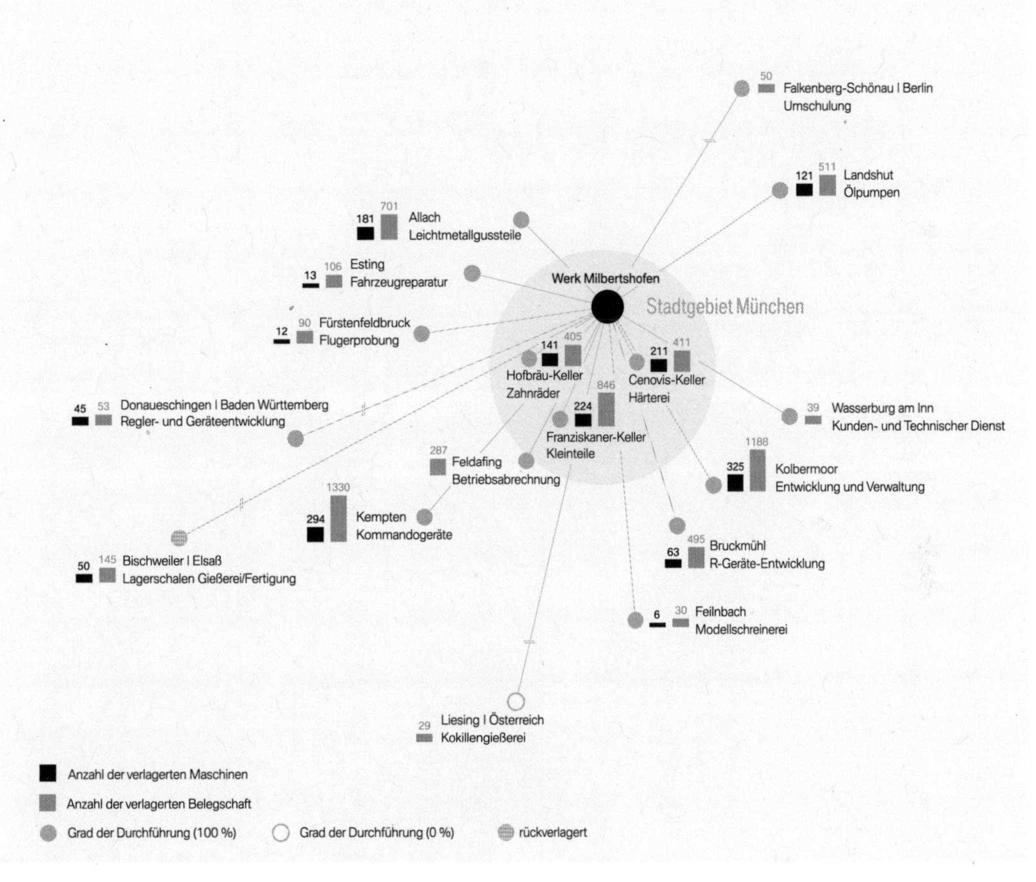

Plan of the relocation in the area around Munich.

the commercial associates were tied up with relocation-related administrative tasks, causing production interruptions that lasted for weeks. The machine pool at the Milbertshofen plant was completely changed. Especially towards the end of the war, communication problems between the various dispersal sites and head office occurred with increasing frequency.

Some of the relocations were within Munich. For example, BMW used space in the Hofbräu Keller as well as in the Franziskaner Keller. BMW tried to ensure that materials were moved to sites within a distance of around 70 kilometres from Munich. The development department was split among three sites. Two sites were near Rosenheim to the south of Munich, namely Kolbermoor and Bruckmühl, while the third was in Donaueschingen, situated some 250 kilometres from Munich. Administration was also moved to Kolbermoor, while aircraft production was transferred to Fürstenfeldbruck. Like the development department, large-scale production and trial production were divided among several sites: Feilnbach, Landshut, Kempten and Bitschweiler in Alsace.

1943 Plans for relocation first drawn up

1944 Relocation of production

Documenting the bomb damage in 1943.

Typing pool at the BMW plant, c.1942.

The workforce

The outbreak of the Second World War ushered in many changes for the workforce at the Munich plant. In 1939 the labour shortage in Munich and the surrounding area worsened dramatically. Not only skilled workers but also unskilled labourers and trainees were in short supply. Even during the war BMW continued its vocational training. An advanced training scheme was set up to train skilled workers, and trainees acquired the necessary theoretical knowledge at BMW's own works school.

As early as the mid-1930s, BMW had changed from a normal 40-hour week to a three-shift operation. Within this model a 65-hour week was not unusual for BMW associates. The labour shortage became even more acute when eligible men were called up for service at the start of the war. However, because BMW was an important armament manufacturer for the Luftwaffe, the company was initially able to benefit from its privileged position: arms manufacturers were provided with workers to replace those called up for military service. In addition, BMW had the possibility of issuing reserve occupation certificates, exempting essential workers from military service. So as to at least partly offset the labour shortage, BMW asked the Munich employment offices to assign female workers to its production plants. But even this failed to meet labour requirements. In 1940 the Milbertshofen plant started using forced labour.

The associates were responsible for safeguarding the production resources during bombing raids on the plants. Secretaries, for example, were expected to take their typewriters with them to the air raid shelter.

The locked-up Herbitus test bed, 1943.

On the production line it was the responsibility of the Italian forced labourers to move the machines to safety – a duty that often cost them their lives. Intensified bombing of the Munich area resulted in a collapse of the infrastructure, and a large proportion of the workforce had no option but to walk to work.

In the course of the war, working conditions progressively worsened. Workers responsible for the Herbitus high-altitude test bed, for example, had to work mainly at night. This innovative facility allowed engineers to simulate the temperatures and air densities of various altitudes in order to test aircraft engines under realistic conditions. However, the system's power consumption was so great that, had it been run during the day, it would have significantly disrupted the supply of electrical power to the entire northern district of Munich.

The plant management tried to avoid plant downtime and meet production targets by initiating numerous accident prevention and cost-saving

Poster: "The end of the alert means straight back to work!", 1943.

Entrance to an air-raid shelter at the Munich-Milbertshofen plant, 1943.

schemes. With resources becoming increasingly scarce, associates were awarded bonuses for their "savings suggestions". Events like the annual Christmas party, at which associates' children received presents, were meant to help forge close links between the workers and their workplace.

In the course of the war, government orders for BMW aero engines steadily increased. The required production boost could only be met by using foreign workers, forced labourers and concentration camp prisoners. Forced labourers were first used in the two Munich plants Milbertshofen and Allach at the end of 1940. Their ranks rapidly swelled from around 500 to some 1,960 in 1941. The percentage of forced labourers in the workforce also grew steadily. By the end of 1941 there were some 4,300 foreign workers and 4,900 German workers at the Munich plants. BMW's management tried to get mainly skilled workers from western Europe assigned to the plants in order to maintain production quality to as high a standard as possible.

At the end of 1943 BMW saw yet another fresh influx of forced labourers. In the meantime, the situation had changed. After their home country had surrendered to the Allies, Italian workers were regarded as war adversaries and as a result they were treated much more harshly. In addition, BMW no longer relied passively on the allocation of workers but began sending company representatives to the Ukraine to negotiate directly with the labour administration there and have workers allocated to BMW. The proportion of foreign workers grew steadily during the war, surpassing the 30% mark in 1942.

The forced labourers at BMW comprised several groups. A large proportion of the forced labourers used in Munich came from western Europe. In the spring of 1943 they were broken down among the following nationalities: 38.7% French (including prisoners of war, otherwise 29.7%), Dutch, Italians, Belgians and Luxembourgers. Poles, Russians and Ukrainians accounted for around 20% of the forced labourers. When BMW first began to use forced labourers, the management made efforts to integrate them in the workforce. This is reflected in measures such as the use of interpreters, camp spokesmen and offices set up to deal with foreigners' general questions and work-related problems. According to the 1943 Social Performance Report, considerable effort was put into carrying out integration measures. However, since this was an official report, the statements in it have to be taken with a grain of salt.

As the situation began to deteriorate in 1942, the working and living conditions of the forced labourers also grew steadily worse. Moreover, the

Propaganda at the south gate of BMW's Milbertshofen plant, 1942.

Forced labourers manufacturing BMW 801 control units, 1943.

rights accorded to foreign workers – at least on paper – applied only to western Europeans, not to Poles and "eastern workers". Low wages and additional expenses meant that, in effect, the Poles, Ukrainians and Russians received no wages at all. With regard to punitive measures for sabotage attempts, absence without leave etc., the forced labourers were subject in principle to the same sanction system as German associates. It appears that in the increasingly chaotic situation that prevailed after the end of 1944, the established hierarchies began to crumble. For example, from the autumn of 1944 associates no longer had to wear "affiliation insignia", e.g. to designate their nationality.

1941 saw the first initiative on the part of the BMW board of management to use prisoners from the Dachau concentration camp. However, the allocation of prisoners was halted because requirements could be met from other quarters. The use of concentration camp inmates was thought to be dangerous because of their opposition to the regime. Nevertheless, the idea was taken up again as early as 1942, when it became more and more difficult to secure workers. BMW and the Reich Ministry of Aviation pursued the same objective in this respect. In 1942, State Secretary Erhard Milch of the Reich Ministry of Aviation asked the SS about the use of concentration camp prisoners in aircraft and engine production. Concentration camp prisoners were also used from 1943 in final assembly, though always in closed gangs who were kept under constant surveillance by the SS and had to work in separate production halls. Physical assaults on concentration camp prisoners were far more common than on

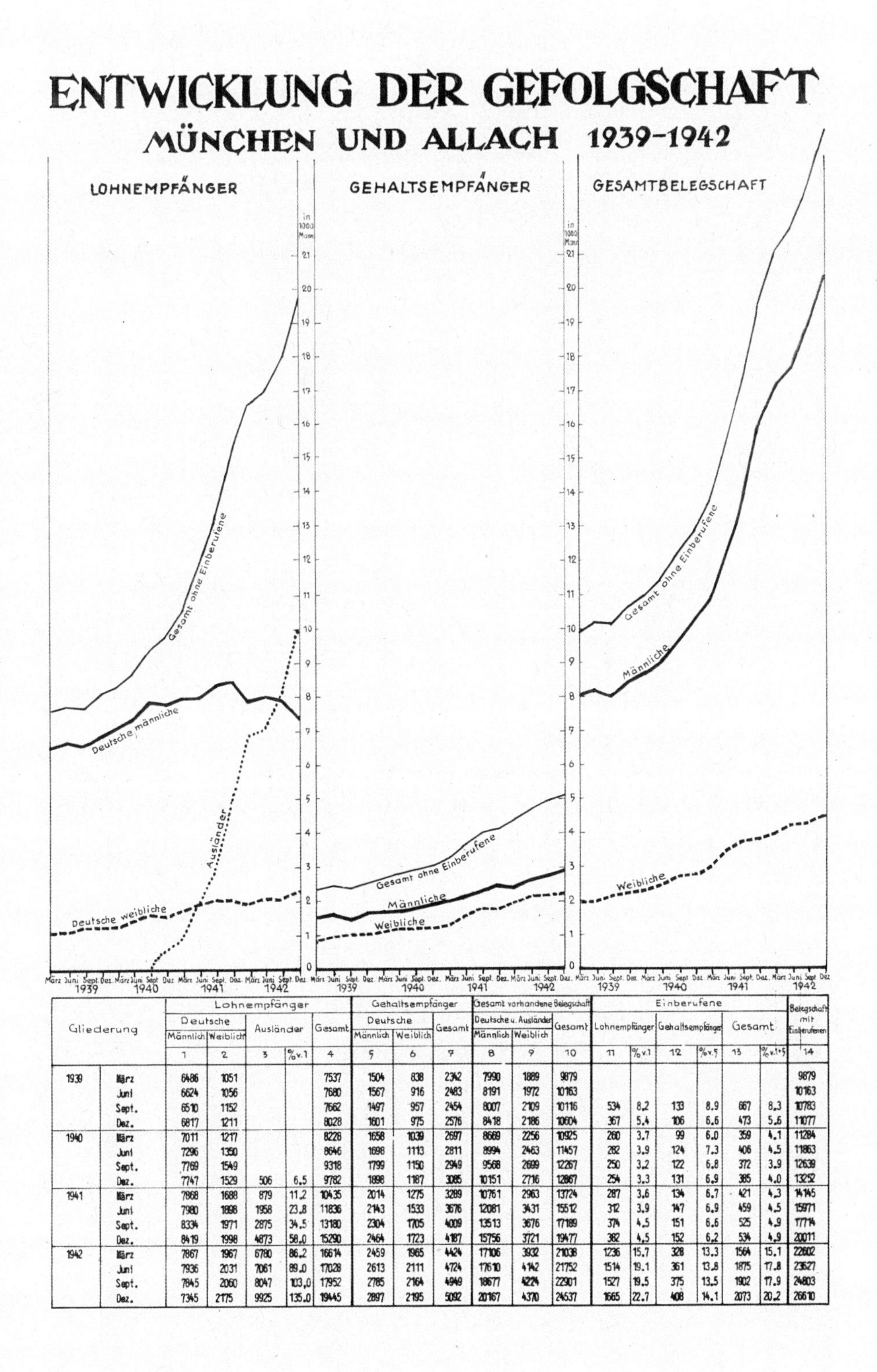

Gliederung		Lohnempfänger: Deutsche Männlich	Lohnempfänger: Deutsche Weiblich	Lohnempfänger: Ausländer	Lohnempfänger: Ausländer %v.1	Lohnempfänger: Gesamt	Gehaltsempfänger: Deutsche Männlich	Gehaltsempfänger: Deutsche Weiblich	Gehaltsempfänger: Gesamt	Gesamt vorhandene Belegschaft: Deutsche u. Ausländer Männlich	Gesamt vorhandene Belegschaft: Deutsche u. Ausländer Weiblich	Gesamt vorhandene Belegschaft: Gesamt	Einberufene: Lohnempfänger	Einberufene: Lohnempfänger %v.1	Einberufene: Gehaltsempfänger	Einberufene: Gehaltsempfänger %v.5	Einberufene: Gesamt	Einberufene: Gesamt %v.1+5	Belegschaft mit Einberufenen
		1	2	3	%v.1	4	5	6	7	8	9	10	11	%v.1	12	%v.5	13	%v.1+5	14
1939	März	6486	1051			7537	1504	838	2342	7990	1889	9879							9879
	Juni	6624	1056			7680	1567	916	2483	8191	1972	10163							10163
	Sept.	6510	1152			7662	1497	957	2454	8007	2109	10116	534	8.2	133	8.9	667	8.3	10783
	Dez.	6817	1211			8028	1601	975	2576	8418	2186	10604	367	5.4	106	6.6	473	5.6	11077
1940	März	7011	1217			8228	1658	1039	2697	8669	2256	10925	260	3.7	99	6.0	359	4.1	11284
	Juni	7296	1350			8646	1698	1113	2811	8994	2463	11457	282	3.9	124	7.3	406	4.5	11863
	Sept.	7769	1549			9318	1799	1150	2949	9568	2699	12267	250	3.2	122	6.8	372	3.9	12639
	Dez.	7747	1529	506	6.5	9782	1898	1187	3085	10151	2716	12867	254	3.3	131	6.9	385	4.0	13252
1941	März	7868	1688	879	11.2	10435	2014	1275	3289	10761	2963	13724	287	3.6	134	6.7	421	4.3	14145
	Juni	7980	1898	1958	23.8	11836	2143	1533	3676	12081	3431	15512	312	3.9	147	6.9	459	4.5	15971
	Sept.	8334	1971	2875	34.5	13180	2304	1705	4009	13513	3676	17189	374	4.5	151	6.6	525	4.9	17714
	Dez.	8419	1998	4873	58.0	15290	2464	1723	4187	15756	3721	19477	382	4.5	152	6.2	534	4.9	20011
1942	März	7867	1967	6780	86.2	16614	2459	1965	4424	17106	3932	21038	1236	15.7	328	13.3	1564	15.1	22602
	Juni	7936	2031	7061	89.0	17028	2613	2111	4724	17610	4142	21752	1514	19.1	361	13.8	1875	17.8	23627
	Sept.	7845	2060	8047	103,0	17952	2785	2164	4949	18677	4224	22901	1527	19.5	375	13.5	1902	17.9	24803
	Dez.	7345	2175	9925	135.0	19445	2897	2195	5092	20167	4370	24537	1665	22.7	408	14.1	2073	20.2	26610

Graphs showing the growth of the workforce in Munich and Allach, taken from the wartime performance report, 1943.

Barrack block housing SS prisoners, 1942.

non-interned forced labourers, though they probably tended to occur outside the plant in the prison camps run by the SS. Treatment of the workers often depended on the character and compassion of the foreman. Living conditions in the Allach subsidiary camp were anything but humane in terms of hygiene, nutrition and general treatment. Whether BMW ever intervened to improve these conditions is not known. Interventions did occur in Munich but only with regard to nutrition, and probably mainly with a view to maintaining the prisoners' work performance.
Especially during the relocation activities from the middle of 1944, the aim of which was to move production underground, the situation for concentration camp prisoners and other forced labourers grew ever harsher, and the already poor working and living conditions became positively inhumane.

1940 Use of forced labourers in production

1942 Use of concentration camp prisoners in production

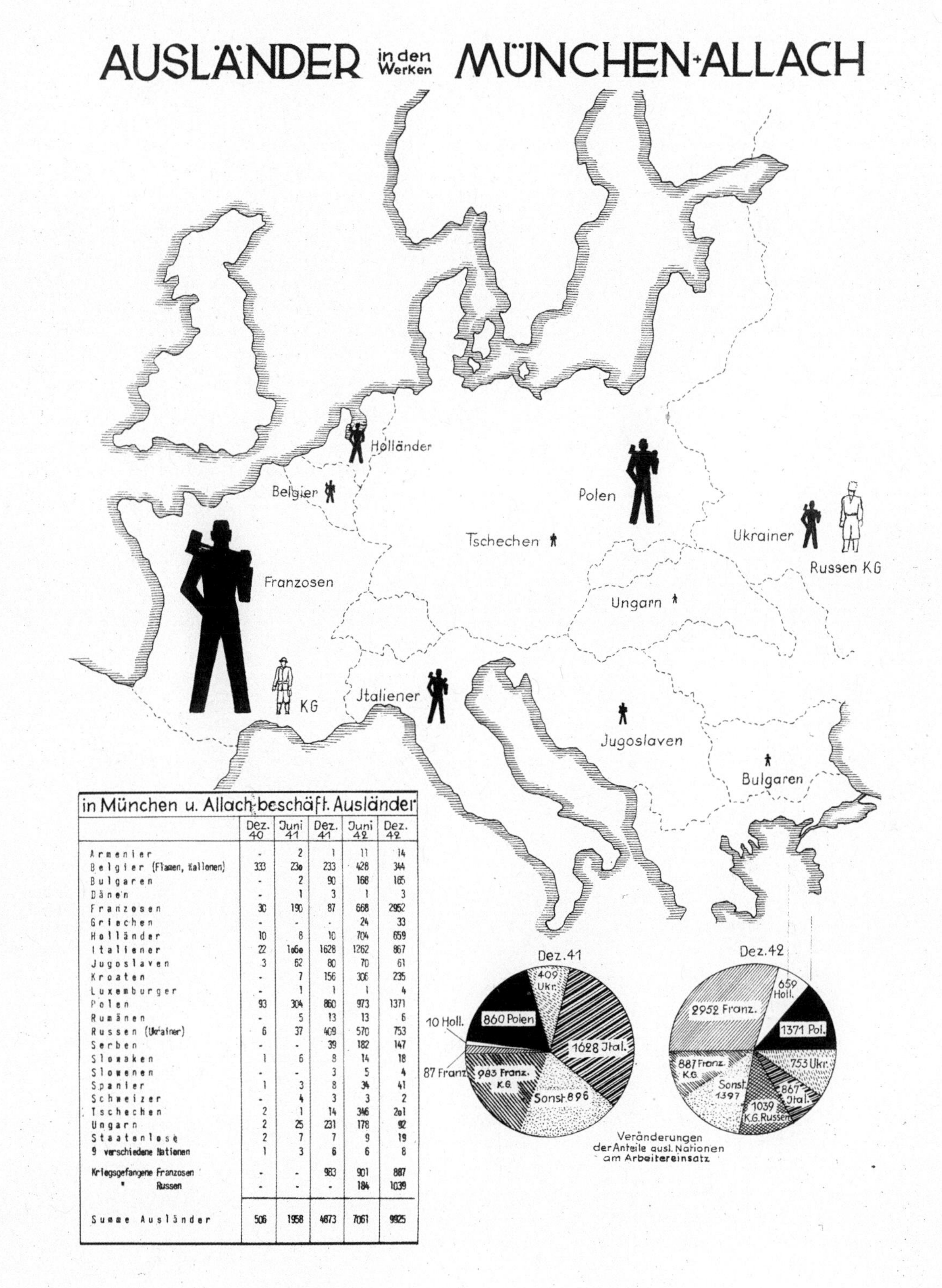

in München u. Allach beschäft. Ausländer

	Dez. 40	Juni 41	Dez. 41	Juni 42	Dez. 42
Armenier	-	2	1	11	14
Belgier (Flamen, Wallonen)	333	230	233	428	344
Bulgaren	-	2	90	168	165
Dänen	-	1	3	1	3
Franzosen	30	190	87	668	2952
Griechen	-	-	-	24	33
Holländer	10	8	10	704	659
Italiener	22	1060	1628	1262	867
Jugoslaven	3	62	80	70	61
Kroaten	-	7	156	306	235
Luxemburger	-	1	1	1	4
Polen	93	304	860	973	1371
Rumänen	-	5	13	13	6
Russen (Ukrainer)	6	37	409	570	753
Serben	-	-	39	182	147
Slowaken	1	6	8	14	18
Slowenen	-	-	3	5	4
Spanier	1	3	8	34	41
Schweizer	-	4	3	3	2
Tschechen	2	1	14	346	201
Ungarn	2	25	231	178	92
Staatenlose	2	7	7	9	19
9 verschiedene Nationen	1	3	6	6	8
Kriegsgefangene Franzosen	-	-	983	901	887
" Russen	-	-	-	184	1039
Summe Ausländer	506	1958	4873	7061	9925

Chart listing foreign workers at the Munich and Allach plants, taken from the wartime performance report, 1943.

TOP
BMW - 163/1/1
GREEK WAR MINISTRY
PIRAEUS
(GREECE)
VIA-BREMEN
BMW
GUSS-ERZEUGNISSE
BMW
FOUNDRY - GOODS
MADE SINCE THE END OF THE WAR

The plant at the end of the Second World War

By the time the German capitulation was signed, BMW's Munich plants were already in the hands of the American troops. The war had taken a heavy toll on the home plant, and much of the factory and machinery had been destroyed. In October 1945, the US military government issued a seizure order for the Milbertshofen site and ordered its assembly equipment and machinery to be dismantled. The following year, production was resumed at BMW with a range of substitute products, consisting mainly of household articles.

1945 – 1952

Occupation of the Milbertshofen plant

By the end of the Second World War, many buildings and much of the assembly line equipment at BMW's home plant had been either completely destroyed or was so badly damaged that it could no longer be used. Motorcycle production had been moved out to Eisenach in 1942, and the entire plant given over to aero engine production. Now, at the end of the war, BMW found itself up against the same problem it had faced 30 years previously: it only had one product to its name, and the market for that product had completely collapsed. What's more, it seemed unlikely that the Allies' ban on arms manufacture would be lifted any time soon. The most pressing question for BMW for the time being, however, was not so much how to return to civilian production but how and whether the company would even survive. American troops had occupied the Milbertshofen plant, the company headquarters of BMW AG, on 30 April 1945. In the four weeks that followed, the plant managers were strictly forbidden to leave the site. During this time Milbertshofen was cut off even from the one other BMW plant in the American-occupied zone, Allach. When the American restrictions on the Milbertshofen plant were lifted again, BMW's management sought permission to resume limited production, with the result that on 28 July 1945 the company was granted a limited permit to carry out machinery repairs and to repair US military vehicles. This permit also authorised the manufacture of spare parts. On 29 August 1945, the American military government further extended the terms of the permit to cover motorcycles, agricultural implements and household articles.

Above: Bombed out production hall at the BMW Milbertshofen plant, 1945.

Below:The Milbertshofen plant at the end of the war, 1945..

A department for "Vehicle Repairs and Factory Transport" was now set up, which even in July 1945 was already providing employment for 88 people. This work mainly involved repairing US army vehicles. Other vehicles, such as tractors and trucks, had to be turned away due to lack of space and personnel. The main difficulty faced by the repair team was obtaining appropriate spare parts, particularly since many of the vehicles were not BMW products. Often BMW workers first had to build special tools before they could carry out repairs and servicing on these vehicles. But despite the difficulties, BMW was determined to continue and if possible expand this line of work.

It was not possible to contemplate any further activities for the time being, however, because on 1 October 1945 the Allied Control Council seized the entire assets of BMW and put the company in trusteeship. Since BMW was

Drawing of a model showing the condition of the BMW Munich-Milbertshofen plant in 1945.

a former armaments manufacturer, its plant and equipment were to be dismantled.

Some seizures had already been announced on the radio in late September, but Allach was the only BMW plant mentioned. On 2 October, however, the management in Milbertshofen learnt that orders had been issued to seize the home plant too.

The trustee appointed for BMW AG was Professor Karl Hencky, building director in the Munich city authority. Hencky now had responsibility for all further arrangements relating to the dismantling of the Milbertshofen

Representatives of the victorious Allies at the Milbertshofen plant, 1945.

plant. He announced the termination of all contracts of employment and the cessation of all work at the plant. As a result, he said, all former rights of the supervisory board, the board of management and the plant management had likewise terminated. To assist with an orderly dismantling, Hencky appointed Kurt Donath, Heinrich Krafft von Dellmensingen and Otto Lampertsdorfer as acting technical, commercial and human resources managers for both the Munich plant and the company. This management team's main priority now was to safeguard the future of the Milbertshofen plant. They therefore approached the Munich city authority and, since they fully appreciated that Hencky too was simply carrying out orders, agreed to work with and support him. This ensured that they were directly informed about all measures and actions and given a say in decisions. While attempting to save as much machinery and equipment as possible, the plant management also helped to organise the dismantling work. To strengthen their standing vis-à-vis the other BMW plants and the dispersal plants, the acting plant managers were now officially confirmed in their positions, though they continued to report to the trustee.

At this time a big question mark still hung over the future not only of the Milbertshofen plant but also of BMW AG itself. For example there was still no clear idea how to proceed on the question of wartime debts or payment of wages. Regarding occupation-era wage arrears and debts to suppliers, except those relating to the Russian-occupied zone, a decision was not reached until the spring of 1946.

The function of the trustee was confined simply to managing the assets of the company, rather than giving any thought to a resumption of peacetime production. In November 1945 there was even talk of handing the plant over to the United Bavarian Cooperatives, as a way of securing its future.

30.04.1945 American troops occupy the Milbertshofen plant
28.07.1945 A production permit is granted
01.10.1945 The plant is seized by the American military authorities

Awtowelo company signage, plus BMW symbol, over the entrance to the BMW Eisenach plant, 1949.

The fate of the other BMW plants

During the war, BMW was operating six German plants. The plants in the Soviet-occupied zone – at Eisenach, Dürrerhof, Basdorf and Zühlsdorf – were occupied by the Red Army and expropriated without compensation. The plant in Spandau, Berlin, found itself in the British zone, though it was a while before communication links were restored with the headquarters in Munich. Here too the acting management of the Spandau plant, appointed by the Spandau municipal authorities, resumed post-war production with a substitute product range. In 1948, BMW Maschinenfabrik Spandau GmbH was formed. The Allach plant, like the main plant, was occupied by American troops, who established the Karlsfeld Ordnance Depot on this site. The KOD became the largest maintenance and repair centre for US military vehicles in Europe.

A dismantled machine tool is removed, 1947.

30.04.1945	Dismantling order issued for the Munich plant
1947	Around 50% of machinery already dismantled

Dismantling programme at BMW's home plant

Above: Preparing to transport a test rig facility, 1947.

Below: Dismantled and crated equipment is taken away, 1947.

The dismantling order issued by the military government in Munich stated that work should begin on the dismantling and demolition of BMW's Munich plants with immediate effect from 1 October 1945. The plant and equipment was to be crated up for shipment out of the country. Only former BMW associates who had not been members of the NSDAP and who had already been vetted by the military government were allowed to take part in this work. The wages these associates received for dismantling tasks were the same as they had been paid for their regular work.

Following the dismantling order, all work at the Milbertshofen plant ceased immediately, including the repair work on US army vehicles. This led to American complaints, however, and on the recommendation of the Transportation Corps, work on American vehicles and the manufacture of spare parts was allowed to resume.

In all decisions as to which items of plant and machinery should be dismantled, the trustee showed exceptional good faith towards BMW. As far as was compatible with his duties, he tried to ensure that at least some parts of the factory were saved for BMW. In an attempt to minimise the extent of the dismantling, BMW had declared around 70% of the machinery to be suitable for peacetime production. Before the physical dismantling of the machines could begin, an official inventory had to be carried out. Due to various complicating factors, this was easier said than done. For example, BMW associates did not have access to all buildings on the site and some buildings were also being used by the US army. To compound the problems, there was also a shortage of personnel who were adequately qualified to perform an inventory count. In the end, dismantling work had to be provisionally interrupted so that the workers taking part could be reassigned to inventory duties. In late 1945, some 600 members of the workforce were deployed on inventory and dismantling tasks. This allowed the count to be completed by the end of the year at both the Munich plants and also at the various dispersal plants. In the final analysis, BMW was required to surrender 9,500 different items of plant and machinery with a total weight of 13,379 tonnes, including 4,200 machine tools.

Now that the inventory had been completed, dismantling was ramped up to full speed again. By early 1947, around 50% of all machine tools had been dismantled and packed up in wooden crates ready for shipment to 16 countries. However, most of the plant's surviving production equipment was never used again. Not until July 1949 did the Allies end their dismantling of German firms, by which time 2,598 machines, with a total market value of 9,245,302 reichsmarks, had been removed from the Milbertshofen plant. The wages for the dismantling work were paid not by BMW but by the central committee of the reparations offices. In addition to the removal of physical property, the most grievous loss was that of intellectual property in the form of designs and engineering drawings. Representatives of all the Allied powers visited BMW's home plant to "harvest" technological advances and developments for use in their respective countries. On 17 October 1945, for example, a British commission visited the plant to look through the development records. The British team arrived unannounced and demanded to see all documentation relating to BMW's R 75 and R 31 motorcycles. When plant representatives initially told them that these records had been moved to Eisenach when production was transferred, the British said that they were wilfully withholding information, and threatened them with punishment. The commission ended up seizing design drawings of all BMW models, including original drawings of the R 255/256 racing motorcycle. Since they allowed no copies to be made, the information was irrevocably lost to BMW. A further major blow was the loss of the Herbitus high-altitude test facility, which was unique of its kind.

Rail wagons loaded with dismantled machinery at the Milbertshofen plant, 1947.

Substitute production

Following the appointment of a new trustee, Hans Karl von Mangoldt-Reibold, in March 1946, the Milbertshofen plant made new efforts to restart civilian production. The idea was that dismantling and reconstruction would proceed in parallel. On 25 March 1946, the military government for Bavaria issued the company with a new production permit. Unlike the first permit of 1945, however, this one only covered the production of household articles and spare parts and did not extend to bicycle or motorcycle production. However, since reparations shipments of dismantled machines were not due to start until the beginning of 1947, the plant management was given permission to use these machines if necessary in the resumption of peacetime production. In addition to household articles,

BMW cooking pots stacked up in a production shop, 1946.

Products	Quantity	Sales value (Reichsmarks)
Machinery and component repairs	200	200,000
Cooking pots	34,000	401,650
Potato crushers	350	1,750
Egg whisks	5,100	20,500
Construction fittings	160,000	72,150
Compressors for Südbremse	225	21,400
Compressor spares	7,000	8,000
Dough dividers	25	15,000
Multipurpose implements	10	2,500
Spares for agricultural machinery	2,000 kg	2,000
Miscellaneous items	n.a.	125,000
Light-alloy castings	175,000 kg	577,000
Grey iron castings	7,000 kg	6,300
Bearing shell linings	3,600 kg	55,400
Spares for BMW cars	19,000 kg	190,000
Total		**1,699,150**

Articles produced at BMW's Milbertshofen plant from June 1945 to May 1947.

including some 34,000 cooking pots, the plant also began the licensed production and sale of the Raussendorf multipurpose agricultural implement. The substitute products were built on the eastern side of the factory site, since the rest of the plant was still being used by American troops. Substitute production continued until the Americans withdrew from the plant in 1952. .

1946 Production permit for household articles and spare parts

1952 End of substitute production and American withdrawal

BMW
25000

Return to motorcycle and car production

In the first years after the war, BMW's Munich-Milbertshofen plant relied on substitute products to keep its workforce busy. Motorcycle production began again in 1948, and the first post-war BMW car came off the line in 1952. Unfortunately, however, these first cars misjudged the mood of the market and sales were slow. In 1955, BMW tried a different tack, entering into a licence agreement to build the "Isetta motocoupé". Even so, by the late 1950s the company was in crisis, and in 1959 there was even talk of a sell-off to Daimler-Benz.

1950 – 1960

Rebuilding the plant

In autumn 1947, BMW received permission from the American military government to prepare for production of motorcycles up to 250 cc. For the time being, permission was still not given to prepare for car production, although the signals were positive. The company management for its part had already decided in 1946 to go back to building motorcycles in Munich. A large part of the former data and records relating to motorcycle development and production had been lost, however, as a result of the war. Either they were in Eisenach, now behind the Iron Curtain, or else they had been seized by the Allies after the war. The motorcycle production lines had gone too, having been moved to Eisenach in 1942.
But a resumption of production faced other problems as well, such as the lack of suitable on-site capacity. The production shops that were still standing after the war proved to be too small. As a first step to resuming operations, the war-damaged production shops had to be rebuilt and huge quantities of rubble removed. Only by 1947 was all debris finally cleared away. By 1949, BMW had invested some six million marks into generating new production capacity and a further 1.5 million marks in machinery and equipment for a return to motorcycle manufacturing. By 1950, investment in reconstruction had already totalled some 18 million marks. BMW still had access to its wartime profits, however, and was therefore able to finance two-thirds of this spending from its own resources.
The shortage of production equipment worsened in 1947 as dismantling gathered pace. In 1949, the Federal German government passed a law to compensate the affected firms. Machines were to be allocated to the dismantled companies by the machinery compensation office of the Bavarian Economics Ministry. Admittedly, not all companies from which machines were requested complied. And although BMW did receive some equipment under the compensation system, this could sometimes be a mixed blessing: many of the allocated machines were in poor condition or were unsuitable for use in motorcycle production. BMW also obtained some machines on loan. Following the currency reform of June 1948, the company had the opportunity to purchase this machinery outright. Before, under the old reichsmark currency, the owners had often rejected purchase offers. In view of the disproportionately high prices that were now asked, however, the BMW management in most cases withdrew its offers and instead stepped up its efforts to buy new machines in the USA.

Kiosk with BMW posters and notice of job vacancies in Munich, 1948.

Reconstruction costs at Milbertshofen plant	
1946	0.637 million Reichsm
1947	1.848 million Reichsm
01.01.1948–20.06.1948	1.10 million Reichsma
21.06.1948–31.01.1949	2.0 million D-Mark

By the spring of 1949, 619 machines were installed at the Milbertshofen plant. Of these, a quarter were either on loan or had been supplied under the compensation scheme.

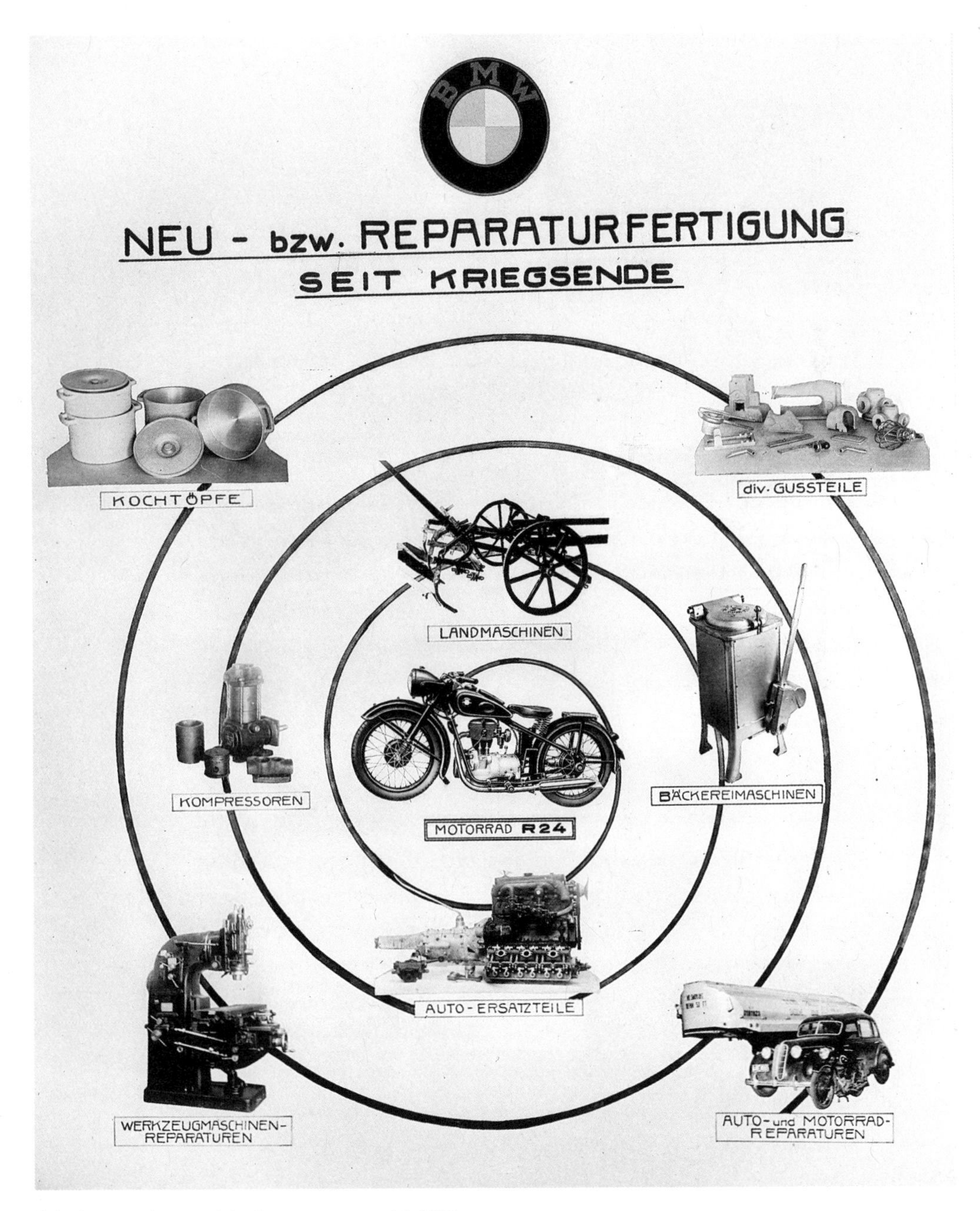

Substitute products and the first post-war model, 1948.

Munich returns to the motorcycle business

BMW's engineering department had started developing a new motorcycle even before the Americans issued a production permit. But only one prototype of this 120 cc machine – the R 10 – was ever built. In the meantime, the plant's technical management had decided on a different strategy, namely to return to the pre-war focus on larger-engined models. Work was hampered, however, by the lack of development and production records. So an R 23 – a 250 cc single-cylinder machine – was stripped down to its component parts and used as a template for new engineering drawings. From these drawings, the engineers then produced new development and production documentation. Soundings taken as early as 1946 had indicated strong demand on the part of the German public authorities for 250 cc motorcycles. But there were no clear indications about private demand prior to the currency reform, so for the time being BMW focused on sales to public customers. Later, when the currency reform in West Germany brought the return of a stable currency system and supply and demand mechanisms returned to normal, BMW was able to gear its production more closely to actual demand. With its R 24, as the R 23-based post-war model was known, BMW had an ideal product with which to make its bid.

Left: The first post-war BMW motorcycle, an R 24, was given away as a raffle prize – to this lucky winner, 1948.

Below: Running at full capacity: the motorcycle production facilities, 1951.

Right: After the war, too, motorcycles were taken around the test track, 1953.

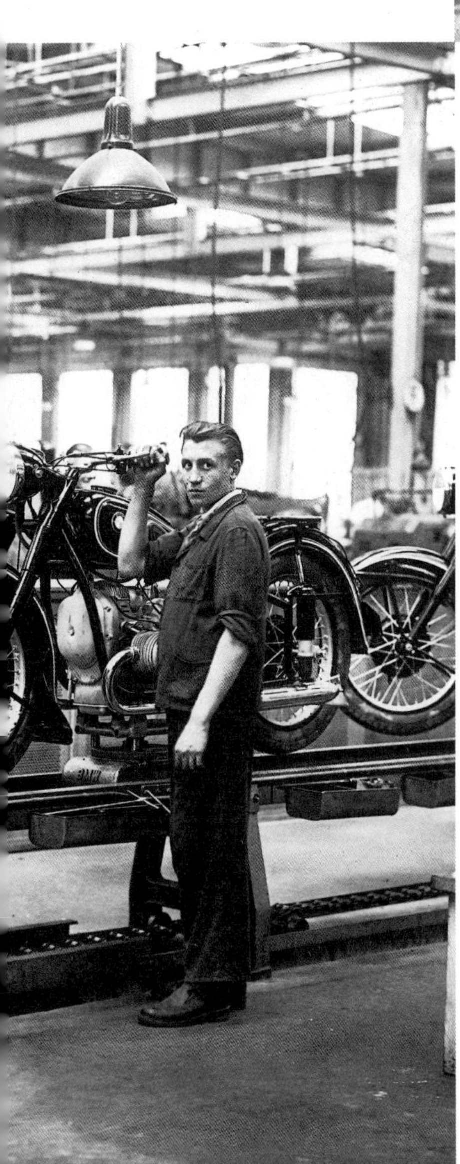

Despite all the above-mentioned difficulties, in spring 1948 BMW was able to present its first post-war motorcycle at the Geneva Motor Show. It was not until 17 December of that year, however, that the first R 24 rolled off the Munich assembly line, due to shortages of materials, plant and equipment. A raffle was then held among the 1,227 associates to find a lucky winner for this first machine. The R 24 soon lived up to all the high hopes that had been pinned on it: production in the first full year, 1949, totalled 9,144 units. Even before the Allies went on to relax the regulations governing civilian motor vehicle manufacture on 13 April 1949, the BMW engineers had begun developing a 500 cc motorcycle, the R 51/2, which was presented to the press in autumn of the same year. Like the R 24, the R 51/2 was based on a pre-war model. This bike too was an instant success: in a production run that lasted just one year, 5,000 units came off the line. The R 51/2 became the most popular German motorcycle over 350 cc with a market share of 90%. Once again, these heavy BMW machines also emerged as the motorcycle of choice for the police. The escort of Federal President Theodor Heuss, for example, took delivery of six of the first bikes to come off the line.

Although the first BMW post-war motorcycles contained no particularly startling technical innovations, these "new" models did help BMW regain its

The BMW R 25/3

The R 25, presented in May 1950, was BMW's first single-cylinder motorcycle with rear suspension. It got a visual facelift with the R 25/2 of 1953, then a complete revision in the R 25/3, which was in production from 1953 to 1956. The most notable changes were a modified fuel tank, full hub brakes, light-alloy wheels and a further improved engine. The official factory output rating for the R 25/3 was 13 hp, though this was widely felt to be an underestimate, given the top speed of 120 km/h. The welded tubular frame with proven straight-travel rear-wheel suspension and four-speed transmission were retained. With total production of 47,700 units, the R 25/3 held onto its crown as the company's most successful motorcycle model of all time right up until the 1990s.

position in the market. Though relatively high-priced, just as they had been before the war, they were also much in demand – and this kept the Munich plant's new motorcycle production lines running at full capacity. The new model range allowed BMW to make the most of the boom which gripped the German motorcycle market during the early post-war years. Since a car was still beyond the means of many Germans, motorcycles became the favourite mode of transport at this time. In the early 1950s, the company extended its product range. In addition to the less expensive "entry-level" single-cylinder models, sports models and touring models, it was now also able to offer prestigious high-end luxury sports models. But by the mid-1950s, the steady upward trend was faltering, with production peaking at just under 30,000 units in 1954 as BMW began to feel the effects of an overall downturn in motorcycle demand. The German "economic miracle" was now getting into its stride, and more and more customers were able to afford the luxury of a car.

Start of car production

Not long after the end of the war, BMW management in Munich started to plan for a return to car production. But there were many problems which first had to be overcome, not least the need for a permit from the American military government. Since BMW car production before the war had been confined exclusively to Eisenach, Munich no longer had access to engineering drawings or plans. Also, in addition to a shortage of appropriate assembly line equipment, skilled labour was in short supply too, since the area in and around Munich had no previous tradition of car manufacturing. On a brighter note, though, some former associates of the Eisenach plant had left the Soviet-occupied zone and decided to bring their skills to the BMW plant in

During the war, models were already being developed for the post-war period, 1940.

Munich. In the course of clearing-up operations during these early days, workers at the Munich plant came across a BMW 332 prototype. Though badly damaged and little more than a shell, the vehicle was promptly reconstructed. With no permit and no production lines to build it on, however, regular production was out of the question.

Despite the Allied production ban, Munich was nevertheless able to use the time to do a certain amount of preparatory work on the development of a new car model. The engineering department favoured a sporty mid-sized vehicle, in keeping with the typical BMW image. Fritz Fiedler, who later became overall head of vehicle development, also designed a small two-seater, the BMW 531, but the board of management decided that a small car did not befit the company image. There would have been a problem on production grounds too: a small car would have required a high-volume

Designs for the first post-war cars in the body development department, 1950.

1952 The BMW 501 goes into production

1955 The BMW 507 is unveiled at the Frankfurt Show

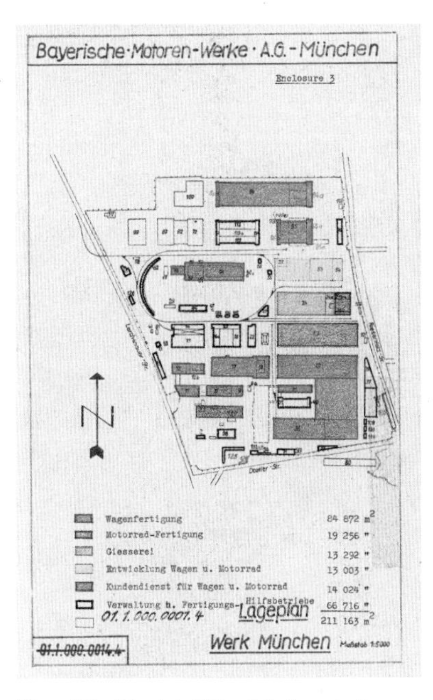

Plan of the Munich-Milbertshofen plant, 1950.

production set-up, which BMW could not have afforded at this time. The board of management therefore resolved that the company should restart car production in the same area of the market where it had left off in 1939, namely the large-car segment.

In developing what was to become the new 501, BMW relied heavily on tried and tested precedents. All design ideas were inspired by the BMW 326, which was in production between 1936 and 1941. A six-cylinder engine was chosen, which had been a BMW hallmark since the early 1930s. The suspension and column-mounted gearshift were borrowed from the reconstructed BMW 332. Due to manpower and tool shortages at the Munich plant, prototype production of the BMW 501 had to be outsourced to coachbuilders Reuter in Stuttgart, who were given strict instructions to have the cars ready in time for the Frankfurt Motor Show in April 1951. This was cutting it rather fine, but BMW was under a great deal of pressure. The company had borrowed heavily to finance the production lines on which the new model would be built, so the sooner they started rolling the better.

Building cars required a lot more capacity than building motorcycles, and the plans drawn up by 1950 envisaged a car production area of 85,000 square metres. A U-shaped complex of buildings in the southeastern corner of the site would house small- and large-panel press shops (Building 33), pre-assembly and bodyshell assembly (Building 32) and the paint shop and upholstery shop (Building 20), while the lavatories and washrooms opposite the canteen in Building 31 would be replaced by electroplating and wheel- and tyre-mounting lines. The final assembly lines would be set up in Building 17. Building 12, to the west of this, was also to be devoted to car production, while in the area inside the test track, the engine test rigs were to be replaced by a test department and a pre-dispatch preparation centre for new vehicles. The car production area was also to comprise two buildings in the area formerly owned by BMW Flugmotorenbau GmbH, in the northern part of the site. These buildings, numbers 95 and 84, were now to be used for raw and finished parts storage.

The BMW 501 went into production in 1952. BMW had high hopes of its first post-war car model, and the production lines were designed for an output of around 20,000 units a year. However, sales of the BMW 501 and its successor the BMW 502, which followed in 1954, fell well short of these expectations. The low sales volumes meant that the production lines had to work well below capacity, and revenues were barely enough to cover the development costs. Nevertheless, BMW decided to add further highly exclusive models to the range. Alongside the saloons, it also wanted to offer a sportier model. The first recorded plans for this only date back to early 1953, but

Slack demand meant production lines for the BMW 501 were working below capacity, 1952.

The large models in the 1956 product range: BMW 507, BMW 503, and BMW 502 (far right).

by 1954 a prototype was already being tested on the Nürburgring. However, the company then heard from the most important export market, America, that the vehicle would be unlikely to sell there in this form. The American importer suggested that BMW should commission the young designer Albrecht Graf Goertz to come up with an alternative solution. Goertz now set to work and designed both the BMW 507 Roadster and the 503 Coupé, which were unveiled to the public at the 1955 Frankfurt Motor Show. But though the BMW 507 went down in history as the "dream car from the Isar", neither it nor the 503 did anything to boost order levels or production, and the production lines in Munich were still only able to operate at part capacity. It was becoming all too clear that, what with the installation of the brand-new production plant in Munich and understaffing in both the development and the production sectors, the company had not had enough time to develop a model policy which would cater to both brand and market requirements.

BMW Isettas on the production line, 1956.

The BMW Isetta

Instead of the two-stroke engine used in the original Iso model, the BMW Isetta was fitted with the single-cylinder four-stroke 250 cc engine used in the R 25 motorcycle. The first version, which went into production in spring 1955, had a displacement of 250 cc and an output of 12 hp. At the end of the year, this model was joined by the slightly more powerful Isetta 300. Externally, these first BMW Isetta 250 and 300 models were very similar in appearance to the original Italian Isetta, although a different bonnet was used and the second, revised model series featured smaller headlamp housings. There were also export versions with large bumpers and large headlamps for the USA, and special versions like the "Tropical Isetta". In late 1956 a revised Isetta with updated body styling was brought out, bearing the name BMW Isetta Export. For a period of several months, the previous model continued to be offered alongside it at a cheaper price, under the name BMW Isetta Standard. The new model, the Export Isetta 250 and 300, was significantly modified compared with the previous version. It went into production in November 1956.

Isettas in the BMW car park at the home plant, 1957.

While in most areas of German industry the "economic miracle" was gathering speed and skilled labour was becoming scarce, Milbertshofen sometimes had capacity standing idle, and workforce numbers had been falling since 1954. Clearly, what BMW needed was a car which would be inexpensive to manufacture and would generate strong sales. But that was easier said than done. The company could no longer afford the whole process of developing a new car from scratch, nor did it have the time. So in 1954 it entered into a licence agreement with the Italian firm Iso to manufacture Iso's Isetta bubble car. A number of design changes were made and soon BMW's assembly lines were turning out the "BMW Isetta Motocoupé". Even though the Isetta made little difference to BMW AG's fraught financial situation, at least it kept the production lines running at capacity and kept the workforce in a job. In 1957 and 1958, workforce numbers even started to rise again slightly. In total, 161,728 BMW Isettas were built between 1955 and 1962.

Above: The Daimler-Benz/Deutsche Bank recapitalisation plan inspired this contemporary caricature, 1959.

Left: Erich Nold addressing the Extraordinary General Meeting, 1959.

Mercedes models from Milbertshofen?

The launch of the Isetta was not enough, however, to sort out the company's financial woes. In 1955 BMW slipped into the red, and stayed there. For a few years, the operating losses were covered by raiding the company's reserves. Obviously that could only be a temporary solution. So Dr Heinrich Richter-Brohm, the new chairman of the board of management appointed in 1957, now prepared a "blueprint for the future" aimed at turning the crisis around. He proposed that, instead of large cars, Milbertshofen should go over to building a "mid-size" car embodying the traditional brand values of sportiness and reliability. In the interim, to boost capacity utilisation, the company would build the BMW 600, a kind of "enlarged Isetta", and the BMW 700. The 700, unlike the Isetta and the BMW 600, was a "proper" car once again.

However, there were still question marks regarding the financing of this new product range. For the time being, therefore, only the BMW 700 went into production, and not the planned "mid-size" car. By autumn 1959 the financial position of BMW AG was so dire that an alliance with a financially strong partner seemed the only way forward. What's more, it looked as if this would take the form of a sale to Daimler-Benz AG. The Stuttgart company indicated that it could place a major order with the Munich plant right away if a deal were agreed. But although such a solution would certainly have rescued BMW, it would have reduced its role, and that of the

A consignment of BMW 600 models leaves the Milbertshofen plant for the customs department in the Federal Ministry of Finance, 1958.

Milbertshofen plant, to that of a mere supplier to Daimler-Benz. It was decided to put a vote on the sale to the Extraordinary General Meeting called on 9 December 1959. Following a nine-hour debate, the small shareholders and the dealers proposed adjourning the meeting and Daimler-Benz' takeover offer expired. However, although the sale to Stuttgart was now off the table, BMW's problems were still far from over.

The BMW workforce

One of the problems facing the Milbertshofen plant when it resumed motorcycle and car production was the shortage of skilled workers. When motorcycle production was restarted in 1948, the associates who had previously been building substitute products were now retrained in motorcycle assembly and engine and component production. Of course, it helped that at this time BMW was still able to call on the services of many skilled workers and engineers who had helped the company build motorcycles back in the 1930s. Their experience and commitment ensured a largely smooth resumption of motorcycle production at the Munich factory.

But the problems when car production was resumed were on a different scale altogether. Here, in contrast to motorcycle production, there was no way that the lack of engineering records could be made up for by drawing on the knowledge and experience of the workforce. The reason there were so few workers with the relevant skills was simply that there was no car industry in the Munich area. However, as the production lines began to roll, the workforce at the Munich plant gradually expanded. This continued when the Americans closed the Karlsfeld Ordnance Depot in Allach in 1955 and some of its associates were transferred to Milbertshofen. Most of these new associates were assigned to production of the Isetta. The "Motocoupé" not only kept the production lines moving and for the first time allowed Milbertshofen to experiment with high-volume production, it also provided a good opportunity for the new associates from Allach, as well as workers from the motorcycle sector, to be reskilled for car production. The following statement to journalists by the technical sales planning director, Helmut Bönsch, shows that the company was very aware of the importance of a well-trained workforce: "The value of a car ... depends not just on the design ideas it embodies but also, and critically, on the quality-consciousness of the people who develop and build it. We at BMW are in the fortunate position of having a team of outstanding engineers, production supervisors and operators who have been fostering a quality culture for many years." BMW's skilled workforce was one of the issues raised at the Extraordinary General Meeting that discussed the possible sale to Daimler-Benz. The opponents of the sale pointed out that no one had attempted to put a value on the workforce, so Daimler-Benz would effectively be getting this "thrown in for free".

Although the workforce was well aware that the company's situation in the second half of 1959 was precarious, everyday life at the plant still went on

Above: BMW launched numerous advertising campaigns to recruit skilled workers, 1956.

Right: Taking up production of the Isetta meant some of the skilled workers from Allach could be kept on, 1956.

pretty much as normal. For example, the development department continued to work on its ongoing projects without wasting time worrying about the fact that none of them might ever get beyond the drawing board. At the same time, the works council took the opportunity to remind the company that the crisis had had a significant impact on the workforce and that they should be "rewarded" for their services and their loyalty to the company.

250 000

“New Class” triggers upturn: the plant in the 1960s

Having weathered the 1950s crisis, BMW enjoyed an unexpected – and unprecedented – upswing following the launch of the so-called “New Class” models. From the mid-1960s onwards the Munich plant began to reach the limits of its productive capacity, but there was no scope for expanding the site as it was surrounded on three sides by residential areas and the unused land on its western edge had been earmarked for the 1972 Olympic Games. However, by transferring motorcycle production to Berlin and purchasing Hans Glas GmbH in Dingolfing, the company was able to restructure the Milbertshofen plant by the end of the decade.

1960 – 1970

The BMW 1500 draws the crowds at the 1961 Frankfurt Motor Show.

A new beginning

After the sell-out to Daimler-Benz AG had been blocked at the Extraordinary General Meeting in December 1959, the engineers in Munich started to work flat out on developing a medium-range car. Even though production of the Isetta and the "full-size" car continued until 1962 and 1965 respectively, the main output at the Munich plant in the early 1960s focused above all on the small BMW 700. At last BMW had a model in its portfolio that could continue the sporty reputation of the company's pre-war models.

The result of the development engineers' work on a medium-range car was presented at the 1961 Frankfurt Motor Show in the form of the prototype BMW 1500. This first model in the "New Class" marked BMW's return to the principle of "Freude am Fahren" ("driving pleasure") that has been used by the company as a brand slogan ever since. The models that followed incorporated various improvements and innovations and secured BMW an unexpected degree of success in the market. But before series production of the BMW 1500 could be launched, the plant had to be adapted to the new requirements. The board of management started by appointing Paul Volk, former production chief at Auto-Union in Ingolstadt, as the new plant director on 1 January 1963. Volk already had experience of series production start-ups, and the idea was that he could bring this to bear on the launch of the "New Class" at Milbertshofen.

Backbone of production in the early 1960s: the BMW 700 at the final assembly stage, 1960.

The foundry after modernisation, 1968.

Modernisation of the plant

Even though the damage inflicted during the war had been repaired by the start of the 1960s, most of the equipment at the plant was still antiquated. The remains of the running-in track were still in evidence at the centre of the site, which was also covered by low huts and sheds that no longer served any useful purpose. Nor did the machinery fulfil modern requirements – for example the press shop still contained antiquated manually loaded clearing presses that had been purchased in the USA immediately after the currency reform. There was a lack of measuring equipment, and on the body production line, weld joints were still made manually using unwieldy pincers. The paint shop – once the pride of the Munich plant – had only the most primitive equipment, and paint was applied by hand. The assembly lines were not up to standard either – many of the workplaces were not designed on ergonomic principles and were dominated by manual procedures. Only engine assembly processes were more up to date, using a precursor of the modern transfer line. Similarly, little thought had been given to the environmental impact of production processes. Waste water treatment was inadequate and levels of emissions were high. Modernisation was to prove a costly process in terms of both time and money: between 1962 and 1971 BMW invested a total of some 1.1 billion deutschmarks in upgrading the plant.

The foundry in the early 1960s, before its modernisation.

Innovative major construction projects

Part of this investment went into a number of major construction projects on the Munich site. As early as 1961, work started on Building 140, which housed mechanical production and engine assembly, and four years later, Hall 140.1 was added. Both buildings were constructed on the site of the former running-in track and aero engine test beds. An innovative approach was taken to the planning and realisation of the project, with staff facilities located in the basement and the ground floor housing the materials store, pre-assembly, auxiliary operations and general technical plant. The product assembly areas themselves were situated on the first floor, with materials being supplied directly through openings in the ceiling of the floor below. This novel approach to the layout of a production unit quickly attracted the attention of planners and designers from all over the German automotive industry, and the new principles of materials flow applied in Building 140 became a model for the entire sector. The large-scale press shop built in Building 154 in 1968 was designed along the same lines.

During the construction of the new press shop, a disused tank was attached to the roof braces to test its strength, 1968.

Building 140 long served as a model for plant construction in the motor industry, 1967.

The boiler house for the Munich plant: Building 100, c.1960.

Plant heating

Until the 1950s, the heating for the Munich plant was provided by a number of decentralised boiler houses, each with a "coal cellar" of differing size. But management now decided to create a single, central boiler house, to be situated in Building 100, in the space left by the original Herbitus test rig removed in 1945. The heating system was initially oil-fired, but in the mid-1960s the search began for an alternative energy source. There were various reasons for this: the city of Munich had called for the company to respect an existing regulation that required chimneys to be installed to a height of 60 metres above the 20-metre boiler house. Oil prices had also increased significantly, and at the same time natural gas had been discovered at several locations in Upper Bavaria. The BMW plant therefore signed the first-ever contract for the City of Munich to supply bulk quantities of natural gas. Further large-scale customers soon followed suit.

After "practice sessions" at the plant, Alexander von Falkenhausen set a world record in 1966 in a Brabham BT 7.

The "New Class"

In the run-up to the market launch of the BMW 1500, 35 zero series cars were built in Munich and subjected to exhaustive testing. Then, on 25 August 1962, the first of the new cars rolled off the production line. At last the company had started producing a model that met all the expectations attached to a BMW: sporty, powerful, compact, elegant and exclusive, the "New Class" 1500 gave BMW a foothold in the German mid-range car market. By early 1963, 50 vehicles a day were coming off the line. But the birth of the 1500 was not an easy one. The vehicle also suffered a number of technical problems and was replaced as early as 1964 by its successor, the BMW 1600. Further models in the "New Class", with various different engine specifications, followed up until 1972. All in all, a total of 339,814 vehicles in the series were manufactured. In March 1964, BMW finally closed down its production of large saloons in order to free up more capacity for the "New Class". Only the Bertone 3200 CS Coupé continued to be manufactured for a further year.

The launch of the "New Class" also marked BMW's return to the world of motor racing. The company achieved its first motor sport successes with the BMW 700 RS, which it had specially developed for hill-climb events. Another BMW that was particularly successful was the 1800 TI, winning many races in touring car championships during the course of 1964. Its success prompted the company to enter the world of Formula 2 racing, and on 22 September 1965 at the Hockenheim Ring the head of the development department, Alexander von Falkenhausen, in a BMW F1 2000 – a Brabham BT 7 with a BMW engine – set a world record in the E-Class (up to 2 litres) for the quarter mile and 500 metres from a standing start, reaching a

Finishing a BMW 1500 after dipping, 1963.

The BMW 1500

The launch of series production of the BMW 1500 ushered in a new era for BMW. The automotive press hailed it as "at last a real BMW". With its simple, elegant bodywork unspoilt by modish gimmicks, the car harked back to the sporty refinement of pre-war BMW models. Alexander von Falkenhausen, racing driver and engine designer, was responsible for developing its four-cylinder 1.5-litre in-line engine with overhead camshaft that delivered 80 hp. But despite its sporty design the car also offered a high degree of comfort, thanks to its front McPherson struts and semi-trailing link rear axle. A total of 23,807 BMW 1500s were built between 1962 and 1964 before the model was replaced by the BMW 1600.

1962 The first BMW 1500 rolls off the assembly line

1964 The BMW 1600 is launched

speed of 220 km/h within a few seconds. The company's return to motor racing also provided plenty of entertainment for the workforce at the Munich plant. At this point in time BMW did not have its own circuit, so testing had to take place on roads within the site – much to the delight of the workers. Eventually, though, the works council and board of management banned such practices.

The "02 Series"

During the celebrations on 7 March 1966 at the Munich State Opera to mark the 50th anniversary of BMW AG, the new two-door BMW 1600 was presented to the public for the first time. In order to mark this model out from the four-door version with the same engine size, it was marketed as the BMW 1600-2 and, from 1971 onwards, as the BMW 1602. This new compact class of vehicle extended BMW's model range by adding a model below the "New Class", but the link between the two series was unmistakable: the chassis and engine had been largely taken over from the BMW 1600, but the bodywork – despite borrowing certain features from the "New Class" – had been adapted by the design engineers to meet the requirements of a two-door vehicle. As with the other series, the "02 Series" offered various models with different engine sizes and outputs. Between 1966 and 1977 a total of 861,940 "02 Series" cars were produced in Munich.

Left: The first BMW 1500 demonstration cars are collected by dealers, 1962.

Right: BMW 02 car bodies in the paint shop, 1968.

Bottlenecks

The unusually high demand for "New Class" and "02 Series" vehicles soon began to cause production bottlenecks. Expansion of the plant was not an option, as it was surrounded on three sides by housing. On its western edge was the empty green space of the former "Oberwiesenfeld" but this, too, was earmarked for another project: the city of Munich had been awarded the 1972 Olympic Games and was soon to set about constructing sports facilities and a stadium on the site, renaming it the "Olympiapark". The Olympic village to house the athletes was built on the other side of Lerchenauer Straße, a mere 50 metres from the BMW plant. During the Games themselves this caused few problems as the plant was closed for the summer vacation, but the village was also intended to serve as normal housing after the Games were over. Unfortunately its proximity to the plant meant that the buildings infringed regulations requiring residential and industrial areas to be no closer than 400 metres apart, and managers feared the emissions inevitably generated by production processes would lead to complaints. If that had been the case, the only relevant issue would have been whether or not pollution or nuisance could be proved – irrespective of who had been established first on the site. The Milbertshofen plant therefore subsequently had to invest large sums in environmental measures aimed at pre-empting possible complaints. At the same time this early (by German standards) introduction of environmental measures gave the Milbertshofen plant a competitive edge in the medium term. And BMW AG was also later able to use the experience gained at Milbertshofen on its other sites.

A new site

As the Milbertshofen plant could not be expanded, another solution had to be found to the problem of how to increase production capacity. In 1966, BMW AG was offered an opportunity to acquire the Lower Bavarian vehicle manufacturer Glas, which would enable it to increase its production capacity relatively quickly. An intermediary role during the negotiations was played by Georg "Schorsch" Meier, who had become a dealer for BMW and Glas following his career as a BMW racing driver. The top management bodies at BMW were divided on whether the sale should go ahead. The supervisory board was concerned that – only six years

Aerial view of BMW's Milbertshofen plant, bordered on three sides by residential areas, 1966.

Administrative building of the former Glas plant in Dingolfing, 1960. / Administrative building of the former Glas plant in Dingolfing with BMW signage, c.1968.

after the company had been restructured – the financial burden would prove too great for BMW. The board of management, on the other hand, saw this as an opportunity to add some 4,000 trained staff to the BMW workforce and at the same time gain access to the market for small mid-range cars.

After carefully weighing up the pros and cons, both bodies finally decided to approve the acquisition of Hans Glas GmbH. As well as the actual cost of the purchase, there were further costs resulting from the need to adapt the plant in Dingolfing to the requirements of BMW AG. The company was not in a position to raise the necessary cash from its own resources as it had already invested heavily in the Munich plant, and the main shareholders rejected the idea of a capital increase, so the acquisition had to be financed through borrowing. The importance of Glas GmbH for the economy of Lower Bavaria was such that BMW received a loan of 50 million marks from the Bavarian government. On 25 November 1966, some five months after the negotiations had first started, the agreement for the transfer of shares was finally signed. Selling price: ten million marks. BMW initially continued production of Glas vehicles, but a combination of problems with quality and a slump in sales resulting from Germany's first post-war recession finally prompted the company to cease manufacturing Glas products in 1969.

Final assembly of the Glas Isar T 700 built in Dingolfing, 1959.

Within a few years of the takeover of Hans Glas GmbH it became clear that the capacity available at the former Glas plant in Dingolfing would not be sufficient for the planned expansion of production. By September 1966, monthly production at the Milbertshofen plant had exceeded the 8,000 mark, and within two years daily output was 500 units. Capacity at Milbertshofen was 100% utilised, and the waiting list for delivery of a new BMW was now some eight months. Company managers therefore decided to build a new vehicle plant on a greenfield site in Dingolfing, and on 22 November 1973 this new facility was officially opened.

Hans Glas GmbH

This agricultural machinery manufacturer was founded in 1883 in Pilsting, near Dingolfing. For many years the main focus of the family-owned company was on repairing – and later developing and manufacturing – farming machinery. In 1907, expansion of the company's product range made it necessary to transfer the plant to Dingolfing. It was only in the 1950s, under the leadership of Hans Glas, that the company started to manufacture motor scooters and cars, and even then it continued to produce agricultural machines. In May 1951 the first "Goggo" scooter was presented in Dingolfing, and by the late 1950s the tiny "Goggomobil" car had also been introduced. The onset of the German "economic miracle" in the mid-1950s also increased the scope for selling larger motor vehicles, and Glas expanded its product range accordingly. However, the company did not have much success in establishing these new models on the market, and towards the second half of the 1960s its financial situation steadily worsened. In 1966 negotiations began with BMW about a possible sale of the company's plants.

Glas Goggomobil, c.1960.

The 10,000th motorcycle is exported to Switzerland, 1953.

Motorcycle production

The decline of the motorcycle industry that had begun in the mid-1950s continued in the decade that followed. BMW was not spared this crisis, but management resolved to continue motorcycle production, albeit at a reduced level. Sales to public bodies, in particular, guaranteed a certain continuity of orders and, combined with exports, prevented a further

Delivery of 250 R 27s to the Bereitschaftspolizei (mobile police), 1964.

decline even when the market failed to recover. Between 1957 and 1969 BMW's annual production figures fluctuated between 5,000 and 10,000, thanks mainly to exports to France and Switzerland – and increasingly also to the USA.

With the success of the "New Class", however, motorcycle production at BMW increasingly began to be treated as a "stepchild", and its importance within the company dwindled. It became obvious that the company's technical and financial resources – and above all the skills of the workforce – could be much more profitably deployed on building cars. But the importance of the motorcycle segment was not completely forgotten. In 1948 it had been motorcycles that had revived the BMW brand and ensured the company's survival, and they were also largely responsible for its sporty image. So BMW continued to manufacture motorcycles, even though it put little investment into updating the existing range, let alone into new production plant. At the same time, the lack of space at the Milbertshofen plant was becoming increasingly acute, prompting management to consider relocating the motorcycle production, which had been based at the home plant since 1923. The board of management was undecided as to whether to opt for the newly acquired plant in Dingolfing or the largely empty site that the company owned in Spandau, Berlin. The decision was finally made in favour of the latter, and since 1967 most BMW motorcycles have been manufactured in Berlin.

Poster showing the BMW R 69 S, 1964.

The BMW R 69 S

In 1960, despite the fact that the motorcycle market had been in crisis since 1955, BMW presented a portfolio of models that – if not new – had at least been revised and updated. One of the highlights was the R 69 S, the fastest ever series-produced German motorcycle, with a top speed of 175 km/h. As the name suggests, it was a further development of the top-of-the-range R 69, with engine performance boosted from 35 to 42 hp. One innovation was a hydraulic steering damper that could be turned on or off. Together with the R 69's tried and tested double swing arm suspension, this guaranteed good directional stability even at high speeds. Various other technical details had been improved by the BMW engineers in their efforts to achieve perfection. Aimed above all at the export market, the R 69 S restored BMW's reputation as Europe's leading motorcycle manufacturer.

The workforce

Up to this point, the development of the company and plant had only called for relatively rough-and-ready workforce structures, and there had been no need for human resources management in the modern sense of the word. The workforce was small enough for everybody to know each other, and communication distances were short. Associate motivation was high – everyone had survived the crisis years and come out on top – and a feeling of solidarity was even shared by those associates who had only joined the company after the major crisis of the 1959 Extraordinary General Meeting. The launch of the "New Class", however, not only marked an upturn in the company's financial fortunes but also engendered changes for the workforce. Especially from 1965 onwards, the nature of the work carried out on the assembly line changed considerably. Following the production launch of the "02 Series", work processes at the Munich plant were rationalised and broken down into small, simple steps. This not only brought them into line with current manufacturing trends but also, by adapting to the new need for semi-skilled workers, acknowledged the need to adopt a new approach to business operations.
While much of German industry had been facing a serious lack of labour since 1955, and companies were increasingly recruiting "guest workers" from neighbouring European countries, this development only started at BMW in the early 1960s, when the company began to recover financially. Initially, BMW tried to attract weekly commuters from the surrounding Munich region, and staff from the personnel department spent countless weekends attending recruitment events in village hostelries. However, it quickly became clear that it would not be possible to cover the company's recruitment needs from the Munich region alone, and BMW therefore decided to start recruiting guest workers as well. The search for suitable foreign workers sometimes took members of the personnel department on highly adventurous journeys – mainly to Turkey, Greece and Yugoslavia, where they would hold job interviews under the southern sun armed with a couple of toolboxes. The new recruits travelled by train to Munich, where they were welcomed at the main station and offered a hearty Bavarian breakfast of roast chicken washed down by a glass of beer. They were accommodated in hostels originally constructed for weekly commuters and were initially employed in semi-skilled jobs. From the very outset, BMW offered them annual contracts – an approach that was by no means usual in the industry. The company also tried to win the

Processing vehicle orders during production of the BMW 700, 1961.

Order details are attached to the car body, 1961.

loyalty of those associates who displayed commitment to BMW, with the result that many of them remained on the workforce even after the recruitment of guest workers stopped in 1973. In a bid to tackle the language barrier, a so-called "BMW learning centre" was also set up, offering a simple, comprehensible introduction to basic terms and concepts related to the immediate working environment and the work techniques involved.

Prayer rooms for Muslims – the biggest religious group among incoming workers – were set up at the plant as early as the 1970s. Individuals from more than 50 different nations were now employed at the plant, and one factor that undoubtedly contributed towards their integration was the fact that workers of Turkish, Greek and Yugoslav origin were elected to the works council at an early stage, and so were able to look after the interests of their fellow countrymen. Associates quickly developed a pride in working for BMW, especially at the company's home plant. Many of them can hardly be referred to as "guest workers" any longer, as they not infrequently celebrate the 25th or 40th anniversary of their arrival. In these early days, information leaflets and circulars were written in several languages and even the company newsletter was multilingual. One reason why integration into the microcosm of the BMW plant was so successful

Today BMW still runs apartment blocks for its associates.

was that it was based on mutual respect and acceptance, particularly of foreign colleagues. In more than four decades the various conflicts occurring between ethnic and national groups – for example the war in former Yugoslavia – have never spilled over into the plant.

Hostels and apartment blocks

BMW hostels were built from the 1960s onwards so that associates from further afield in the surrounding district could have a roof over their heads during the week. Housing has always been in short supply in Munich and the hostels – since converted into apartment blocks – also offered initial accommodation to many foreign associates. Today they are mainly inhabited by interns or associates who have transferred to Munich from elsewhere. Guest accommodation is also available for associates from other company sites who come to Munich for short-term projects and those who have been transferred to the city from foreign sites.

BMW
1 MILLION

From the early 1970s to the mid-1980s

Throughout its long history the Munich plant has seen a number of radical changes. Many of these were introduced during the decade from the mid-1970s through to 1986/87, and their effects are still tangible today. Urgently required space was created by relocating the company headquarters to a new building outside the plant premises. Munich was the birthplace of the company's most successful model series ever, the BMW 3 Series. A new plant at Dingolfing was built in order to keep pace with an ever-growing demand for BMW cars. Larger vehicles were now assembled at the plant in Lower Bavaria. An unparalleled drive towards modernisation revolutionised production, beginning with the body shop.

New company headquarters in Munich, a new plant at Dingolfing

In 1972, just in time to coincide with the Summer Olympic Games held in Munich, a spectacular building went up directly adjacent to the various sporting arenas of the Olympic Park which themselves were the focus of keen international attention: the new BMW AG administrative tower at Petuelring. Built to accommodate about 2,000 associates, the four-cylinder design created space for all key corporate functions. Prior to this, the Munich plant had been bursting at the seams. Producing over 180,000 vehicles

Since the mid-1970s all available land around the plant had been built on.

each year and about the same number of engines, it was the largest production facility for cars and the only one for engines. It also accommodated almost all key corporate functions, including development, sales and human resources. In the early 1970s the only office space outside the actual plant premises was in nearby Building 80 to the south of Dostlerstraße. With the construction of the Olympic village, the Parkhaus West multi-storey car park and the new Group headquarters with its low-rise

The Munich plant in 1967, looking north. Within a few years the area to the west would accommodate a new car park, the Olympic village and the Olympiazentrum underground station. The area to the south was still largely parkland.

The BMW Group plant Dingolfing in an aerial view (2016).

satellite building and the Museum, all remaining gaps in the neighbouring development were now plugged and clearly defined the plant perimeter: Dostlerstraße to the south, Lerchenauer Straße to the west, Preußenstraße, Pommernstraße and Lüneburger Straße to the north, and Riesenfeldstraße to the east. Around Lüneburger Straße, residential housing had encroached almost to the factory gates. If it had not already done so, the Munich plant had now lost its greenfield aspect once and for all.

Along with the tangible relief from space shortage brought by the construction of the tower, the year 1972 ushered in a further turning point for the plant in Milbertshofen. The opening of Plant 2.4 in Dingolfing meant that car assembly could now be launched at a second production facility; the new plant took over production of the new 5 Series, which had been launched in Munich only a few months earlier. The additional relief for the home plant provided by the Dingolfing plant came not a day too soon. Amid the automotive boom of 1973, vehicle production at the Munich plant increased to over 193,000 units – more than three and a half times the figure for 1962, the first full year of production for the "New Class", with which BMW successfully negotiated the crisis point in its automobile business. The Munich workforce had no reason to fear a downturn in status. All the signs pointed towards continued growth.

1972 The new Dingolfing plant takes over production of the new BMW 5 Series

1973 Vehicle production at the Munich plant increases to 193,000 units

BMW defies the crisis: birth of a superstar

By 1974, however, the period of growth was over – at least for the time being. On 17 October 1973, the price of oil went up from around three US dollars a barrel (159 litres) to more than five dollars, and during the year that followed it soared to over 12 dollars. The German government attempted to combat this development, referred to universally as the "oil crisis" or "oil shock", by introducing a series of rather symbolic measures such as a ban on Sunday driving on four weekends in November and December 1973. But the automotive industry was hit particularly hard and suddenly by the consequences of fuel price increases. On 23 April 1974, the German daily newspaper *Süddeutsche Zeitung* wrote: "Throughout April, German car manufacturers have been forced to introduce the most serious production restrictions to date. According to provisional data provided by the dpa news agency, at least 116,500 associates in the car industry (including those employed in Belgian production plants) will be obliged to work shorter hours."

Plant security officers check BMW associates leaving the plant.

Control centre at the Munich plant, Building 18. Seated at the semi-circular 19-metre control wall, a single associate kept a close eye on how the conveyor systems in the plant's key production areas were functioning.

To begin with BMW seemed unaffected by these developments. Vehicle sales were down by just 7%, and some export figures were actually above levels for the previous year. No production restrictions were deemed necessary until well into the autumn. The company announced in its internal associate newspaper *bayernmotor* that it would only be introducing a total of three weeks of shorter working hours at the Munich plant for the period November 1974 to January 1975.

Rudolf Moser
(Plant Security)

I joined BMW AG in Munich as a plant security officer in 1986. BMW provided me with secure employment many years ago; having come from a small factory of just 14 associates, it took me a while to get used to such a large plant.
My new responsibilities were many and varied. To begin with we were given a grey uniform and a peaked cap. In those days there was little in the way of technology at the gates and entrances to the buildings. Visitor passes were written out by hand. Paper was used for most things. Nowadays our work is supported by technology and made a lot easier by turnstiles, cameras and automatic entry systems. Our uniforms have also become much friendlier and help give a more positive image.
The role of plant security has been restructured to some extent over the years. That's why I have always seen it as a personal challenge to be able to change and adapt to new responsibilities. It is great to witness the way the plant and BMW Tower now shine with new splendour.
BMW provided me with secure employment many years ago, and in all those years I have never had to worry about my job.

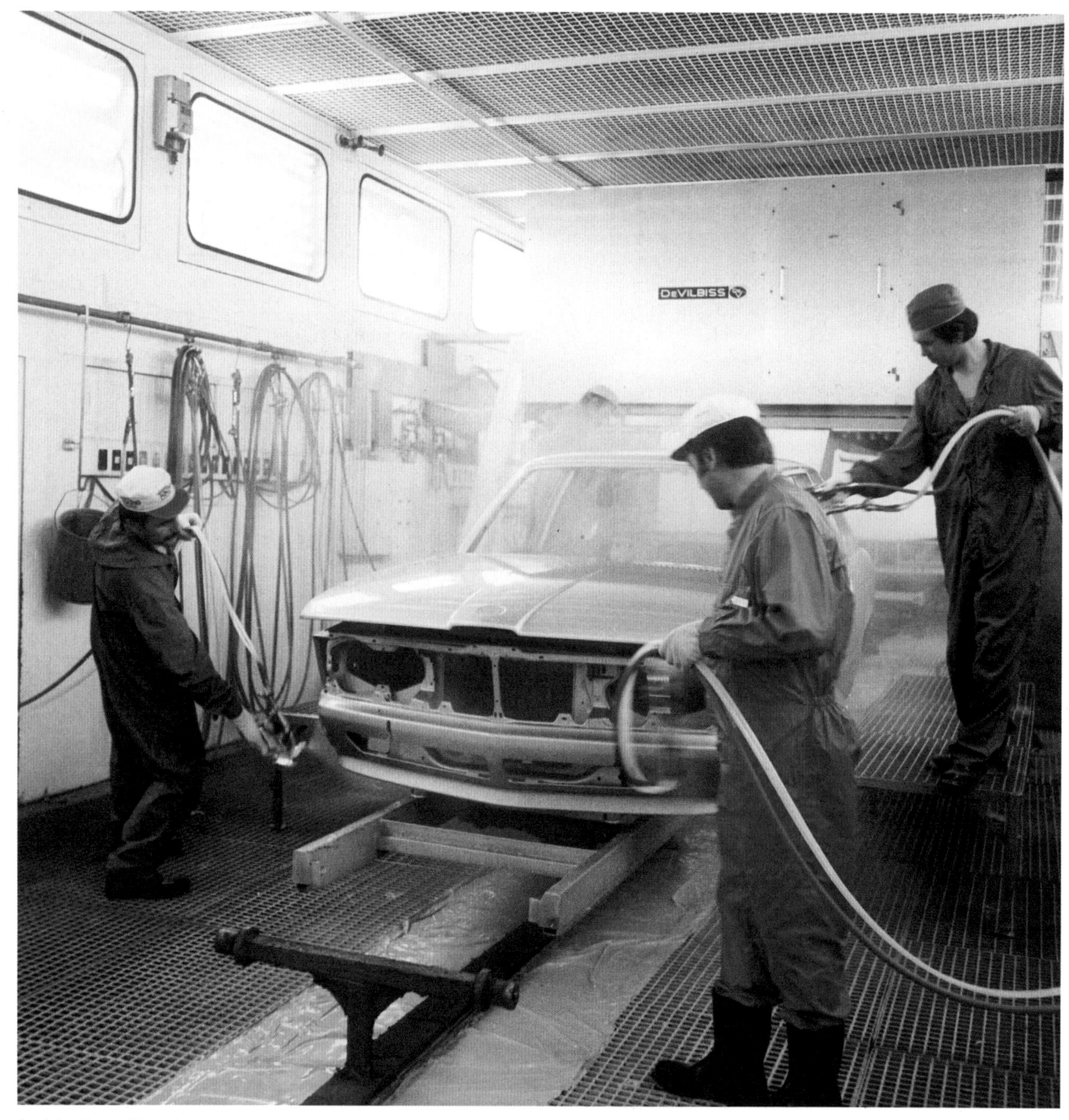

Applying the cavity seal.

1975 Start of production of the BMW 3 Series

Final coat drying.

So was there cause for concern at BMW too? Was this development not similar to the crisis in the late 1950s, when the company landed in difficulties owing to an outmoded product range? After all, sales of the 2500 – 3.3 Li series plummeted to half the level of the previous year and the much-vaunted new 5 Series was now to be moved in phases to the new Dingolfing plant. Yet by 1975 the Munich workforce was already doing extra shifts in order to build 6,000 more vehicles than had been originally planned. Models from the 1502 – 2002 Turbo series proved even more popular in 1974 than the year before, although these did not account for the upturn in sales alone: 1 June 1975 saw the start of production for an all-new series that was groundbreaking both for the plant and for the company as a whole. It was the birth of the BMW 3 Series, the company's most successful series ever.
The company was well aware of the importance of this new generation of sporty, compact saloons. Future developments were to prove this

Final assembly work in the engine compartment.

commentator, writing in *bayernmotor*, right in his assessment: "By giving the green light to the 3 Series, BMW [...] has taken the most decisive step for the years ahead. The position of the company in the second half of the 1970s will not be decided by the 5 Series from Dingolfing alone. The basis of the company's business must come from the new 3 Series from Milbertshofen."

In the period up to the end of that year, over 43,000 first-generation (E21) 3 Series vehicles left the production lines at Milbertshofen. The achievement had required a major effort from all involved: "The Munich BMW team has months of hard work behind it and more ahead. BMWs are now more popular than ever before. The plants at Munich and Dingolfing are together producing 1,080 cars every day. ... In spite of winter knocking at the door, we still have delivery times to keep. ... The first 3 Series production models came off the Milbertshofen production line on 1 June. By late October the plant was delivering 500 of the new models every day. Taking works holidays into account, that means this increase in unit numbers was achieved in under four months. Thousands of works associates have adapted to the new model. The core team was required to train around

End of the assembly line in Hall 17.

1,500 workers who had come new to the company in everything from the basics upwards. Plant manager Paul Volk was therefore 'particularly pleased that a good level of quality had been reached, even though this increase in output had called for exceptional efforts on the part of plant management, foremen and associates.' ... Volk reserved his special praise for the associates 'who, in critical start-up circumstances, were prepared to make personal sacrifices in the interest of completing the task in hand'. Planning for the 3 Series coincided with the oil crisis, the most critical period ever experienced by the automotive industry. ... Business is back on track, unit numbers are once again up." (*bayernmotor* Nov./Dec. issue 1975)

So BMW reached the turning point in a crisis that had left the company virtually unscathed. And since the company rewarded associates for achievement and success already in the 1970s, as a bonus for the year 1975 every BMW associate was given a profit sharing worth 3% of their gross income.' (*bayernmotor*, issue Nov./Dec. 1975)

One of the plant locomotives, photographed in 2003.

Plant and neighbours: not always a harmonious relationship in those days

Within half a generation the area around the plant had changed dramatically. BMW could no longer be portrayed as a factory on the outskirts of the city, as had been the case in the early 1960s. The plant was now a city-centre factory, bordering on a densely populated residential area to the north and east. From now on the term "neighbourhood" was one that could be applied literally.

Increasingly, therefore, the plant was obliged to respect the legitimate wishes of those living in the immediate vicinity. The company and plant management quickly came to the conclusion that the – objectively incorrect – attitude that "we were here first" was not a very defensible position to adopt. Compromises had to be made where production and logistics were concerned, but also in terms of associate conduct. From today's perspective the rather serious tone of contemporary reports published in *bayernmotor* concerning rail traffic between the plant and its external warehouse seems almost comically anecdotal: "Over one hundred railway wagons are picked up and brought every day by the BMW plant locomotive from the Milbertshofen station to the plant and back. 'Apprentices seem to find amusement,' says Otto Weber, responsible for plant transport, 'in running on the rails.' The plant locomotive is therefore obliged to sound its whistle. And that can be heard for miles around. Since October, BMW has been operating a new locomotive. The old, noisy engine now takes the line across the 'green meadow' to Dingolfing. The Munich locomotive still has to sound its whistle from time to time (although it is much quieter). In addition, the engine driver and shunter now communicate by radio."

In particular, local residents had to put up with largely unregulated delivery traffic both inside and outside the plant premises. It might happen, for example, that on Monday mornings the trucks would have to queue up outside the plant, owing to a bottleneck at the unloading area. The new plant gate in Lerchenauer Straße brought a noticeable easing of the situation – the Olympic village, designed for the Summer Games of 1972, had been built allowing a good distance to the through road (Lerchenauer Straße) and the adjoining plant premises. With the new gate in operation, a one-way traffic system was introduced: arriving delivery trucks entered the plant via the northeast gate in Riesenfeldstraße, and after offloading exited via the new gate in Lerchenauer Straße. In order to reduce traffic noise at night, the Riesenfeldstraße gate was closed at 8.00 p.m. During busy delivery times the number of truck journeys into the plant was reduced by building a new warehouse next to the Milbertshofen station. This served as a temporary depot to allow delivery vehicles to be switched increasingly to less busy periods. Parking restrictions imposed on trucks in Riesenfeldstraße proved less successful, however, as they merely resulted in backed-up delivery vehicles being forced to park in neighbouring residential streets.

View towards Lüneburger Straße from the works.

In 1981 trucks were finally stopped from exiting the plant via Riesenfeldstraße (Gate 5), and vehicles used to transport new vehicles were also prohibited from entering through Gate 5. Even cars and small vans could no longer use the exit after 5.00 p.m. As a result, truck journeys in congested Riesenfeldstraße were cut by 200 per day.

In the mid-1970s the company became the first automotive manufacturer in the world to make an appointment to the post of environmental officer. Now BMW could present a human face to the plant's neighbours – Manfred Heller. Heller moved into the neighbourhood himself in order to experience and judge for himself the effects of production and logistics on local residents. It also meant the residents could report their concerns to him directly.

At the same time, it permitted the company to generate a better understanding among residents of the way the plant worked, launching an ongoing dialogue between the two parties.

And yet the plant was (still) limited in its good intentions – by the state of technological development in the paint shop and foundry. Here, at least, gradual improvements to the level of pollutant emissions allowed the plant to keep abreast of what was feasible. What follows is an example dating from 1982, following a comprehensive modernisation of the paint shop (*bayernmotor* 4/1982): " 'We are clearly at the forefront of technological development.' Herbert Felix should know: as head of installation engineering at Munich he is jointly responsible for all systems technology and at the same time plant officer in charge of air pollution control. In 1967 the plant was 'producing' 220,000 cubic metres per hour of waste air from the drying shop. In several steps this volume has been reduced to 55,000 cubic metres – despite greatly increased capacity – all of which is subjected to afterburning. Even more impressive is the reduction of emissions for total carbon [which] today [amounts to] less than 50 milligrams per cubic metre; that represents a reduction since 1968 to 1% of the starting level. 'Unfortunately,' Felix explained, 'it was not possible to reduce the subjective perception of smell by a similar percentage' – the reason being that aromatic substances are detectable even in extremely small concentrations. [...] For this reason Felix considers the idea of emitting waste air via tall chimneys to have little merit. [...] He feels greater success can be expected by developing new paint processes allowing further significant reductions in waste air volumes."

The (punch)cards are shuffled

The success proved the planners right. Only when substantial investments were made, in particular in the body shop, was a new production strategy possible. A world first in body construction was the introduction of the Siemens-designed SIMATIC S30 control module, the first stored program control to replace the logic of wiring by programming using Boolean arithmetic, in other words using and/or links. Programming was done either on site by means of single commands or using cards. Thereafter, data processing took place in the computer centre with the documentation and production of a control tape ready for input. The S30 was able to control two largely automated welding lines for assembly of the underbody and further body framing.

And *bayernmotor* naturally took great pride in reporting this and all the other far-reaching innovations it brought: "One in two BMWs ordered today is a model from the 3 Series. The new series is up and running. The majority of body-in-white construction for the 3 Series is now automated, with just one man supervising the machines where once ten or 15 people were required. The miles of suspended rails, overhead trolleys and points all play their specific roles. For previously – with the construction of the BMW 1502 to 2002 models – almost all bodies were still nameless. These BMWs-to-be only began to develop a life of their own on reaching the paint shop. Today's cars display their individuality at a much earlier stage – already in the body shop they have their own routing cards. And we have realised a long-held wish for the people in the body shop, where associates now travel with the

SIMATIC S30 control module with programming unit and control tape input unit.

apron conveyor instead of walking alongside. Sorting is done overhead, where there is also a materials buffer in case of malfunctions. Parts for body construction can then be called up in the correct sequence. A process computer controls the entire system. It looks for the right parts. ... BMW makes its name from the fact that everyone who wants a car from Munich gets individual service. The earlier we can begin to put the mosaic in place to meet customer requirements, the better our control. With the launch of the 3 Series, BMW has taken a major step forwards in this direction."

Josef Fusseder
(2006, Body Shop)

For me, BMW is a very personal success story. When I started out as an electrician almost 40 years ago there was very little in the way of technical equipment. It was mainly mechanics and hardly any electronics, you could say. Nowadays it's almost all electronics. When I first joined the company I thought I wouldn't be here for long. But you can do anything if you really put your mind to it. At some point I was made a foreman and I've been group leader for a few years now. I'm particularly pleased that I have been able to help with the appointment of many young people. It's nice when you can see what they have done with their lives. As a foreman you need to be a good judge of character. Lots of people have skills you would never have thought possible at first. And that's how you have to deploy your staff. When I look back, I often wonder how we really got where we are today. I well remember the days when we were a small niche manufacturer, when wages went up by penny increments. I was able to afford my first BMW in 1971; I've still got the invoice. And after every new production start-up we all thought: this is the last, after this one our plant will be shut down. Nobody talks about that any more. BMW has built up its image enormously over the years. Now when people ask me 'Where do you work?' and I say 'At BMW,' they all reply: 'Fantastic!'

Quo vadis Munich? Constantly changing emotions

The boom continued at BMW – and before long the plant was again threatening to burst at the seams.
On the one hand, it had become increasingly difficult in view of the lack of on-site expansion possibilities to meet ever-rising demand; in particular, as well as numerous extra production shifts for the plant maintenance team, this meant achieving the impossible and creating extra production capacity during the plant's annual summer holiday. Expanding production capacity was the main goal during the customary four-week break in production; increasingly, too, importance was attached to measures to bring about improvements to environmental protection and to the civilisation of the workplace, two of the desirable side-effects brought about by increased automation. Where body construction was concerned, the central Buildings 32 and 33 grew vertically in order to create room for high-bay storage and integrate a second production level. For the period from the late 1970s to the mid-1980s, the company planned investment in the home plant amounting to over a billion marks.
On the other hand, it was foreseeable that with a realistic – that is to say, reasonably optimistic – assessment of the medium- to long-term order situation, production capacities in Munich would sooner or later run up against more or less insurmountable difficulties. The opening of the Dingolfing plant was a first step towards relieving overstretched facilities in vehicle production. However, Munich remained the company's only production facility for engines until the new engine plant in Steyr, Austria was commissioned in 1982. Both new plants offered the possibility of almost limitless expansion if required. Which explains why the Munich workforce began asking questions about the future of their plant, and why apparently contradictory reports appeared in issues of *bayernmotor* from 1977.

1972 The new company headquarters is built, the new Dingolfing plant is operative

1976 Production of the first generation of BMW3 Series begins

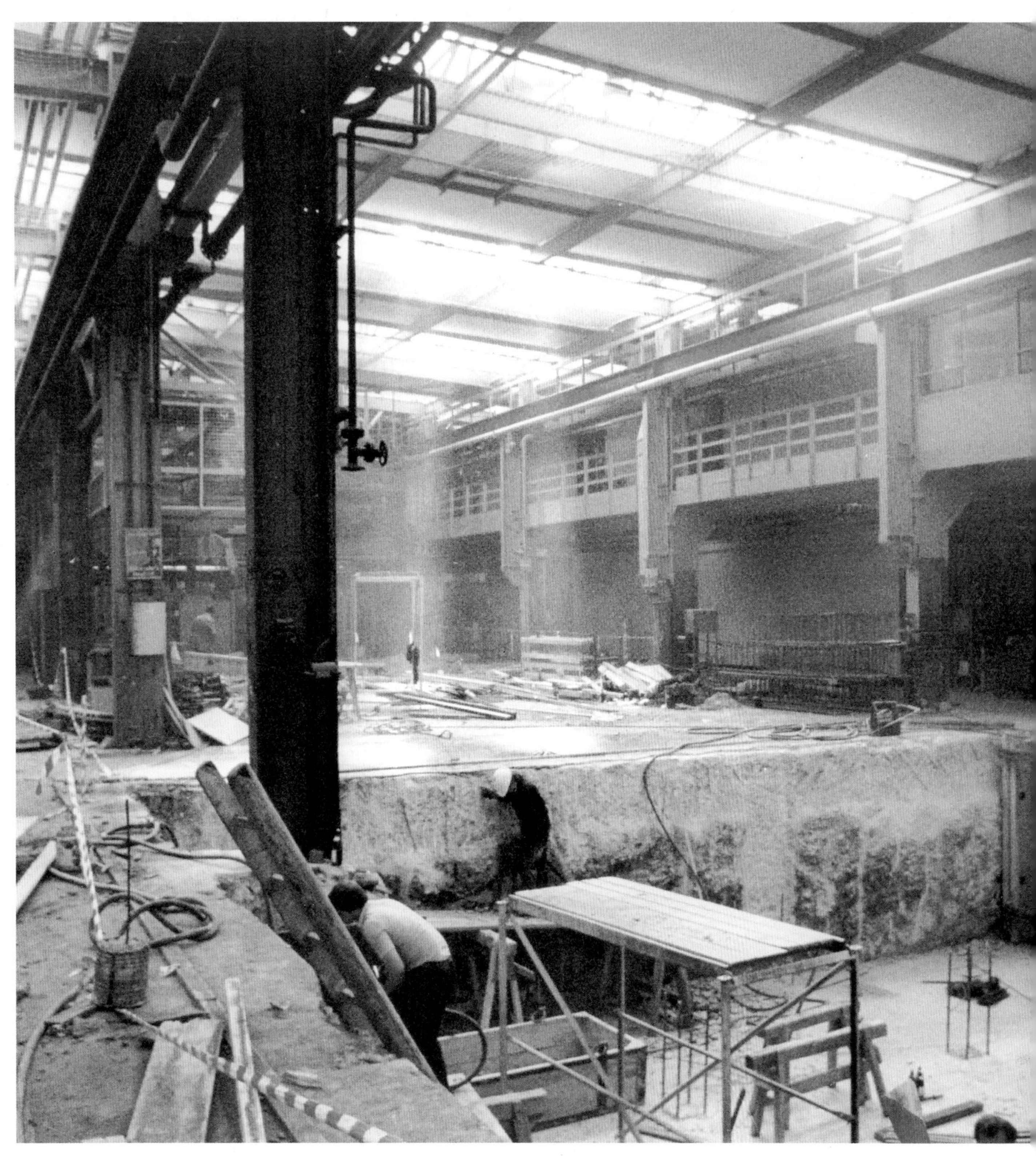

... and production was due to re-start in just a few weeks ...

The first article ran with the headline: "Munich is and remains BMW's main plant".

"Speaking at the associates' meeting on 21 April, Eberhard v. Kuenheim, Chairman of the Board of Management at BMW AG, explicitly stated that BMW Munich would remain BMW's centre and its main plant. 'All rumours,' the Chairman said, 'that there would be cuts in the number of jobs in Munich are without any foundation.'"

Kurt Golda, the chairman of the works council, also spoke out emphatically against rumours of production of the 3 Series moving to Dingolfing. The large 2500 models up to 3.3 litres had been discontinued in February, thereby freeing up much-needed capacity in Hall 16 for expansion of 3 Series production. A quarter of a century after car production started in Munich, the last "large" vehicle left the home plant.

Production of the new 7 Series started at the Dingolfing plant a short while later. While the Munich team played a major support role in the transfer of the 5 Series to the plant in Lower Bavaria, the 7 Series was the first model for which the Dingolfing team could take credit alone for production start-up. As Golda put it: "Long-term planning, whether up to 1982 or 1985, understandably means you bring to the table various considerations for examination and discussion. ... Decisions will not be made before 1979 at the earliest. These will certainly take the line that no jobs in the entire current workforce will be under threat."

Assembly of the large six-cylinder models (E3) in Hall 16. Engine size gave each model its designation (for example, BMW 3.3 L).

The last E3 in the body shop, February 1977. The Munich plant produced a little over 200,000 units in a production run of almost ten years.

Low-pressure die cast technology in the foundry. Liquid metal was cast into the mould from the bottom side. Low-pressure die casting technology was used for large parts like cylinder heads.

Inspecting the crankcase of an in-line six-cylinder engine during the second half of the 1970s.

On the other hand, there were capacity problems (quotation from *bayernmotor*): "The foundry and the machining division of the engine shop have currently reached capacity limits. Many plant areas are having to keep production going for up to 20 hours per day, often in a third shift ... This exceptional state of affairs [...] will be over by the time of the next plant holiday. Then – following the completion of capacity expansion measures – the necessary relief will be felt. ... Preparations have been in progress since the last plant holiday. Production was restructured while the Milbertshofen team was on vacation. [...] In total, BMW is investing well over 100 million marks in expanding engine and parts production."

The engine shop in the mid-1970s produced almost three times as many units as the more conservative planning of the previous decade. Around 1,200 four- and six-cylinder engines left the plant every working day in almost 200 different models and model variants. Capacity was so stretched that engine shop associates did not even have a chance to familiarise themselves with the new layout of production facilities before they went on holiday. The production manager there promised his associates that on their first day back after the holidays nobody need fear he would not find his machine ... During a three-week plant holiday in 1976 there were no fewer than 240 separate construction projects on buildings, supply systems and installations.

Assembly line with four-cylinder engines from the M10 Series.

The new look – more than a fresh front

A new large press shop was completed in 1968/69 and given the building number 154. This shop was urgently needed for capacity expansion: between 1967 and 1970 alone, vehicle production in Munich had doubled to over 160,000 units. The Press Shop still dominates the southwest aspect of the plant. Conversion to fully automated operations began there in 1982; by 1986 the semi-automated "iron hands" that were used to pass parts from one press to another for moulding had been replaced by fully-automated "feeders".

The new press shop at the junction of Dostlerstraße/Lerchenauer Straße, 1969.

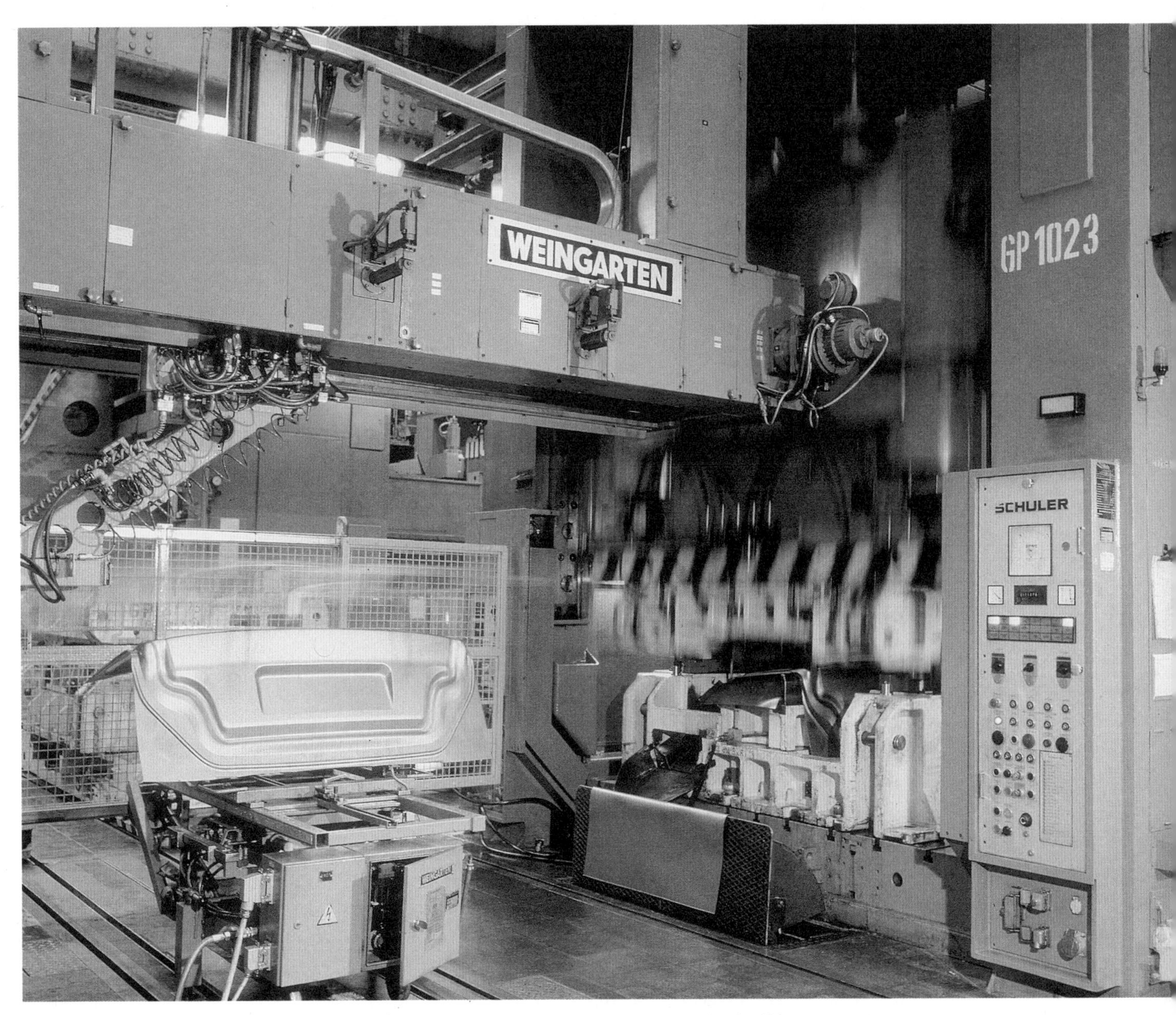

A feeder arm at the Press Shop, pressing line 102-3. This line was operative up to 2011.The photo was taken in 1996.

In 1969 Buildings 158/159 went up in the northwest corner of the plant premises. Today their front aspect is dominated by the fire engine garages built in 1984. From 1980 the northern part of the complex served as a pilot plant preparing for new vehicle production start-ups in Munich and Dingolfing. The pilot plant replaced pilot assembly in Building 95, where since 1970 ten associates had been assembling pilot vehicles.

The complex of buildings 158/159 in 2006.

Initially staffed with around 120 associates, the role of the pilot plant was to test the manufacturing sequences planned for later series production for each new model two years ahead of production start-up. In particular, the purpose of the pilot plant was to help solve one problem: on account of the relatively late active role played by production and process engineering in the development of new models, it was often only possible to identify necessary improvements to the production technology or the product itself in the pre-production series. And this cost a great deal of time and money. So the pilot plant was given the remit to exchange information at an early stage between experts from the development departments and production planning, to design operations planning vehicles in order to identify problems with production engineering, and to train associates from the plants in which the new model would later come off the production lines. Building 159 had three floors at its disposal, each with a usable floor space of 2,000 square metres, for pilot vehicle body construction, pilot vehicle assembly and method workshops.
The pilot plant enabled the company to gain invaluable experience in determining the future structure of the Research and Innovation Centre, which was set up in Knorrstraße in 1986.
From the late 1970s onwards the plant gates were given a more homogeneous appearance, in line with new corporate identity guidelines designed among other things to improve the recognisability of company buildings. This process began with gates in Lerchenauer Straße and

The old main gate from the 1930s. The photograph documents a change of shift, c.1980. The gatehouse was demolished in 1984.

The new Gatehouse 34, Gate 1 in Dostlerstraße.

Riesenfeldstraße, now numbered Gates 3 and 5 respectively; the last gate to be redesigned was Gate 1 in Dostlerstraße. The old plant gate dating from the 1930s was demolished in order to make way for the construction of Gatehouse 34. 1981 saw the demolition of a building that had

Building 11, c.1963. The company address of BMW AG at the time referred to this building (Lerchenauer Straße 76).

dominated the plant's appearance in Lerchenauer Straße for decades, the historic Building 11. Until 1973 the former board of management building had accommodated a small museum at ground-floor level. Then in the 1970s it housed a number of different occupants, the last of which were associates from the body shop, in need of additional office space. Demolition became necessary in order to make way for an additional production building as part of the restructuring of assembly operations.
The endless construction work on buildings and facilities during company holidays provided material for countless anecdotes. In 1978, for example, an electrical short circuit resulted in a premature machinery start-up in the body shop: "Not all assembly lines were at a standstill during the company holiday – at least not in Building 32 in Milbertshofen. Here, during the night of 18 August, 13 (!) painted bodyshells came off the final assembly line – amid a great deal of crashing and banging. The 'ghost line' started up some time between the hours of eight in the evening and six in the morning, when a rat with an appetite for control cable triggered a short circuit. ... The bodies, which owing to 'system filling' had been placed on the conveyor ready for production start-up after the holidays, suffered serious denting." (*bayernmotor*, 9/1978)

1968 Construction of the large press shop number 154

1969 Buildings 158 and 159 are completed

1981 Demolition of Building 11

Herbert Felix
(2006, former Head of Plant Installation)

What was it that fascinated me about 'our' plant and its products?
My love affair with the BMW brand began with my first car, a BMW 700 Sports Coupé. With that car I experienced such intense 'driving pleasure' and pretensions to sporting prowess that the memory of it still uplifts me today.
Only later – when I had gained promotion from BMW driver to BMW associate in 1967 – did I become familiar with the plant as the production facility of such desirable motor cars and was once again infected with the BMW bug.
I started out in the plant engineering department as a 'production and process engineer', my job being to develop and supervise the technical functioning of all systems in the paint and assembly shops. As such I witnessed – and was even part of – an unparalleled advancement within the company in the 1970s. In those days the largely independent yet team-oriented approach often allowed us to realise our ideas and give free rein to our improvisatory skills – which were still an essential part of the job at the time. The market success of cars produced from 'our' installations gave such a boost to us 'non-productive' guys that we thought nothing of enthusiastically 'making ourselves at home' at the plant at night-time, at weekends and, of course, during company holidays.
During those years the sense of a new departure, the tangible rewards for our efforts and demonstrably positive results engendered in me and many of my colleagues an almost intimate and enduring affection for 'our plant' right up to and even after retirement. I'm particularly pleased that nothing ever came of the occasional deliberations to close the plant down for strategic reasons.

SIMATIC S5 control unit for underbody production.

High-tech at the expense of jobs?

Also in 1981 a new building – entirely unspectacular from the outside – was completed to the south of former Building 11, in which the new and greatly improved SIMATIC S5 installation control module with fully automatic welding line celebrated a world premiere. For the first time the installation control module featured microprocessors. Now much smaller in size than before, the control units nevertheless ran into megabytes in terms of memory. In Hall 55 it was now possible for the first time to produce underbodies for several different body variants for one model, including subvariants such as left- and right-hand-drive models, on a single production line and in any sequence. What made this possible was an orange box measuring 12 x 5 cm and known as the Premid (Programmable Remote Identification). Every underbody had a number which was stored in the Premid along with all data required for body construction. Whenever an underbody reached the conveyor belt, the Premid system communicated by microwave with a so-called communicator. This box then issued the command for the next working step. The Premid knew, for example, whether to produce an American model with a special bumper holder or a body prepared for optional equipment features such as a sliding roof.

Underbody production in Hall 55. This production line was in operation until 2004 – for three generations of models. Its robots made 2,430 million spot welds and handled 1,032,240 tonnes of steel in total.

A total of well over 200 robots used in underbody production gave an enormous boost to automation – a development which in view of the constantly growing demand for BMW cars would not result in job losses, as works council chairman Kurt Golda emphasised at a works meeting in May 1981: "We are in favour of the use of robots in jobs which in the long term pose a risk to the health of individual workers. ... To date no jobs have been lost at BMW through automation. On the contrary, jobs are [...] being created. ... A 'no' to modern technology [...] would be potentially to jeopardise not only competitiveness, but the company as a whole – including the jobs of the entire workforce. This is not a line anyone could seriously expect a works council to support."

The picture shows on the right the press shop Building 154 (with logo), in the centre Building 55 with its many office windows, and on the left is the Building 10 complex.

Similar sentiments were expressed by board of management chairman Eberhard v. Kuenheim at the same meeting: "Nobody at BMW need fear for his job on account of the introduction of new robots. As long as we are able – as in the past – to hold our ground in the world market successfully, there will be more than enough work here. ... The times ahead will be harder than we have become used to in recent years. But if each and every one of us knows his job and does what needs to be done, then there should be no really hard times." Nevertheless, the electronic revolution posed enormous challenges for "seasoned" electricians and maintenance staff. The new technologies and processes required an extended running-in period before the team could operate them faultlessly – after all, they had no experience or model from which to learn. Now they were being asked to master electrics, mechanics and electronics all at the same time – a job requirement that had no parallel in any trade at the time. The body shop team had to face up to a challenge that was certainly not made any easier by the simultaneous production start-up of the second-generation 3 Series. Initially they were forced to introduce a third shift at night in addition to normal two-shift production to make up for production downtimes. But the efforts of the production and maintenance staff were not in vain: initial teething troubles were soon overcome, and the production concept of the Munich body shop began its triumphant march. Similar

The BMW Group Rosslyn plant, 2005. In February 2015 the millionth BMW 3 Series was built.

control technology was soon applied to other areas of production and the Dingolfing plant. Five-storey Building 55 – that, too, was a first – housed three production levels and two storage levels. The conveyor bridge (Building 40) to Buildings 32 and 33 could handle 200 underbodies or 60 bodies. Systematic preventive maintenance was now introduced during machinery downtimes in order to guarantee maximum availability of the production facilities and avoid unplanned production stoppages.

Fledglings

No sooner had robots been introduced than increased capacity was once again overstretched – as readers of the March 1983 issue of *bayernmotor* discovered: "An extra shift run between 11 p.m. and 6 a.m. has immediately increased daily unit numbers. In addition, since mid-January about 70 associates from Dingolfing have been lending support to their Munich colleagues in the body shop. Their deployment is scheduled to run until late April and will be seen also as a training period. For it is likely they will all be working with technology similar to that here in Hall 55 in the new body shop in Plant 2.4." In general, the Dingolfing associates seem to

have enjoyed their deployment to Munich. When asked how they got on there, one colleague replied: "Munich makes a pleasant change for me. Discos, cinema, Schwabing – need I say more?" Another worker particularly enjoyed the working environment, which remained friendly despite the capacity overload: "It's always great fun here in Hall 55. I work with a lot of guest workers and we all get along really well."
Deployments such as these have become a routine part of the BMW Group's production network. Not only have they proved invaluable in developing new plants; they also ensure a rapid transfer of knowledge in the spirit of partnership. In particular, many Munich workers spent extended periods working on site when plants were undergoing development at Dingolfing (production start-up 1972), Rosslyn/South Africa (1972) and at the Steyr engine plant in Upper Austria (from 1982).

Helmsman over two decades

For almost 20 years, from 1963 until the end of 1982, Paul Volk was plant manager in Munich. During his time there the home plant was transformed from a workshop operation to a professional and well organised car factory.
The days of "inspired improvisation" (Eberhard v. Kuenheim), which Volk experienced when he joined BMW AG from Auto-Union in 1963, came to an end with the company's rapid growth from the early 1960s onwards. This ended – perhaps with the pendulum swinging too far in the opposite direction, as later corrections showed – in a phase of clear organisational demarcation and highly specialised division of labour. On the other hand, without the restructuring of processes and responsibilities, it would hardly have been possible to increase the plant workforce by a factor of four in the two decades under Volk's stewardship. This restructuring not only involved cooperation between the various plant departments, it also meant cooperating with the development departments, most of which were still housed within the plant premises. This concomitant was a vital factor in the rise of BMW from small car factory to global corporation.
On Paul Volk's retirement from the company, *bayernmotor* took the opportunity of highlighting his personal achievements and the difficult circumstances of his early years:
"His first major challenge was to modernise the production facilities. On his arrival, the plant had around 5,000 associates, and the workstations

Paul Volk

were old-fashioned, inefficient, and by today's standards harsh and inhumane. ... Despite considerable resistance, Paul Volk was responsible for Munich becoming the first passenger car factory in Germany to introduce the innovative technology of electrophoresis. It was his enthusiasm and ability to improvise that got Plant 2.4 in Dingolfing on its feet – whether in terms of planning or the complete relocation of 5 Series production from Munich to the new plant."

Electrophoresis is a paint process closely related in principle to galvanisation. It had been used in Munich for priming since 1966. In electrophoresis the car body is connected to a DC electricity supply in a bath of watery paint. One pole charges the paint particles floating in the electrolyte (the water), which are then attracted to the opposite pole (the body) to form a highly uniform and water-insoluble coat of primer. Any loose or superfluous paint particles are then rinsed off with desalinated water and only the electrophoretically applied coat is subsequently hardened in the continuous paint drier. In this way all surfaces, including cavities, are uniformly primed as a basis for effective corrosion protection and more economical paint coverage.

When asked about working conditions in the plant in the early 1960s, Volk said:

"Here in Munich the advantage lay in the fact that construction work could take place above existing halls while work continued as normal underneath.

In this way the body shop was enlarged in good time, making it possible to switch production from the [small BMW] 700 up to larger series. ... As far as available investment was concerned, things were a bit tight in those days. ... We posed the question: what is most important, what do we want most? And what was always important to us was being faster than others."
In 1963 Paul Volk established the tradition of "foremen meetings", a unique combination of workshop and social get-together. This tradition continues today ("supervisors' day"). One Saturday each autumn the plant management says "thank you" for the year's work and presents an overview of upcoming challenges. The foremen, regarded by the Munich plant as managers, remain enthusiastic participants.

Models and products: two generations of the BMW 3 Series, two new six-cylinders

The first-generation 3 Series (E21), launched in 1975, was a success from the start. By November 1978 the production line discharged its 500,000th 3 Series.
In 1976 Eberhard v. Kuenheim, chairman of the board of management at BMW AG, aired his views on the new 3 Series in a double-page advertisement entitled "BMW is leaving its niche": "This model series has brought success that goes well beyond our expectations. The new vehicles no longer just satisfy the sporty ambitions of a relatively small circle of discerning motorists. They also target a larger group of experienced drivers, for whom quality and safety are more important factors. We therefore appeal now to a much wider market."
BMW scored a hit: just a year after its debut, the BMW 320 was voted best saloon in the sub 2-litre category by readers of Europe's number one specialist car magazine. And that was not all: the new series was the first vehicle in its class to get a six-cylinder engine. When the two new models 320/6 and 323i were unveiled at the Frankfurt Motor Show in 1977 they became the focus of all BMW enthusiasts. The combination of manoeuvrable, sporty

The first-generation BMW 3 Series.

saloon and silky-smooth, comfortable yet muscular six-cylinder was unique in the marketplace. All-round disc brakes supplied the necessary retardation. It meant the 3 Series was not only faster than many higher-classed cars, but also equipped with superior technology.
The popularity of the 3 Series generated untold conjecture about and analyses of the car and its drivers. In one such study from 1980, 77 percent of people surveyed gave performance as the most important purchase criterion, 65 percent cited manoeuvrability, and 64 percent the saloon's sporty appearance. Almost two thirds of respondents said they would certainly buy another BMW when it came to changing cars. And 80 percent were of the opinion that nothing about the 3 Series required improvement. In 1981 an unambiguous figure put such high-level customer satisfaction beyond doubt: in May of that year, almost six years to the day after production start-up, the one millionth 3 Series came off the production line. That

The BMW 320i as four-door variant, available from 1983 on.

made it the most successful BMW of all time. In 1982 BMW introduced the comprehensively revised second-generation 3 Series (E30). In carrying out much precision work on the bestseller, the engineers had pulled off a brilliant coup. For although the new version now offered four centimetres more room inside, the body length was actually three centimetres shorter. The front indicator lamps had been moved from the wings and integrated into the bumper, and the centre of the car now featured a broader, sturdier-

1975 Start of production of the first-generation BMW 3 Series (E21)

1978 The 500,000th 3 Series comes off the production line

1981 The one millionth 3 Series comes off the production line

1982 Start of production of the new 3 Series (E30)

1983 Introduction of the four-door version of the 3 Series

looking, matt-black B-pillar. The wedge shape with high rear end had long since gained general acceptance, and nobody gave a second thought any longer to the relatively large rear lamps. On the contrary, with its extra 35 millimetres of track width the new 3 Series generated a much more powerful presence. It was an impression confirmed when you took to the road in it. In the first place, the BMW developers had ordered a strict diet for the 3 Series, resulting in a weight loss of 30 kilos – despite the inclusion of extra equipment features. Secondly, the broad bonnets now housed some even more potent power units, and in particular – thanks largely to much improved aerodynamics – the new two-door version ran faster. By the end of the first year of production alone, BMW had notched up production figures of 233,781 new 3 Series. Such a popular reception was more than anyone had expected. As a result the Dingolfing plant was obliged also to begin building the 3 Series much earlier than had been anticipated – during the start-up year 1982. And this despite the fact that the real surprise was not due until autumn 1983: the four-door version of the 3 Series. With this BMW had responded to growing demand from potential customers for more comfortable access to the rear seats. Having a family with children was now no longer a reason not to treat oneself to a 3 Series.

With the Evolution, BMW brought out a small special-edition series of even more powerful M3 models. Recognisable by its even more lavish spoilers, the special M3 was fired by a 220 hp engine. And naturally a version with catalytic converter was available from late 1989 that delivered 215 hp.
The first generation of M3 models rolled out of the Munich plant from 1986 onwards. In contrast to other high-performance vehicles, the sporty 3 Series was not hand-assembled in small unit numbers, but produced as a large-volume vehicle on the assembly lines. For the M3, for which a first edition of no fewer than 5,000 units was planned, had to be a road car fit for everyday use. The M3 proved that the recently introduced catalytic converter technology need not necessarily compromise power. With exhaust filter fitted, the sportiest of all the 3 Series still managed a peerless 143 kW or 195 hp (four-cylinder S14; 2.3-litre displacement, four valves). By the end of 1991 the Munich plant had turned out 17,970 M3 models.

F

New structures and first steps towards a “new” Munich plant

The relocation of the remaining development departments to a new site gave the Munich plant an opportunity to restructure – and to reflect on its mission. For example, was it still appropriate for a car plant to be situated in the middle of a major city? Two new plants were built during this period – in Regensburg and in Spartanburg, South Carolina – with the aim of increasing vehicle output and manufacturing additional models and body styles, while the Munich plant now focused even more closely on its core tasks of vehicle and engine production. With the Z1, BMW revisited its low-volume production traditions of the 1950s and started to introduce new work structures.

1985 – 1995

The Research and Innovation Centre approx. 2005, seen from the BMW Tower.The roofs in the foreground belong to the paint shop at the Munich plant.

R & D moves out

Even if physical expansion had been a realistic option, producing new body versions for the expanding 3 Series while at the same time producing engines not only for the 3 Series but also for the likewise booming 5 and 7 Series models from Dingolfing would have been too much for the Munich plant to handle on its own. In fact, even if it had devoted itself exclusively to production of the 3 Series Saloon (which up to the model change in 1990 comprised both four-door and two-door versions), the runaway sales would still have brought capacity problems. The low-volume models – the M3 and Z1 – added to the challenge. To expand engine production capacity, BMW had already set up a new plant in Steyr. And now the company decided that it needed a new car production plant too, in addition to Dingolfing. The choice of location fell on Regensburg.
But it wasn't just production capacity that had to keep pace with expanding vehicle sales. The increasing technological sophistication of the product also required more research and development capacity. Already by the early 1980s, the Munich plant was no longer able to keep pace with this, particularly since vehicle and engine production required more space too. Furthermore, the need for increasingly close cooperation and integration between the development departments could not be served by dispersing them between different buildings, some of them situated off site. The time was ripe for a change of tack, and so the company decided to build a research and engineering centre at a site in Knorrstraße (today known as the Research and Innovation Centre).

Space at last!

The relocation of the development departments gave the Munich plant new opportunities for structural optimisation. By 1983, the new plant manager Jakob Gilliam was already setting out a strategy: "We must try to adapt our jobs and workplaces to the needs of the coming years. Our plans include the following: to alleviate capacity problems in engine production in Hall 140 by using the Steyr plant and by installing new engine test rigs; to transfer large tool manufacturing to the Bundesbahnhalle, which will improve workplace design in the press and tool shops; to alleviate capacity issues in the vehicle assembly shops by integrating them with the buildings

Assembly configuration at the Munich plant before and after the 1984 restructuring

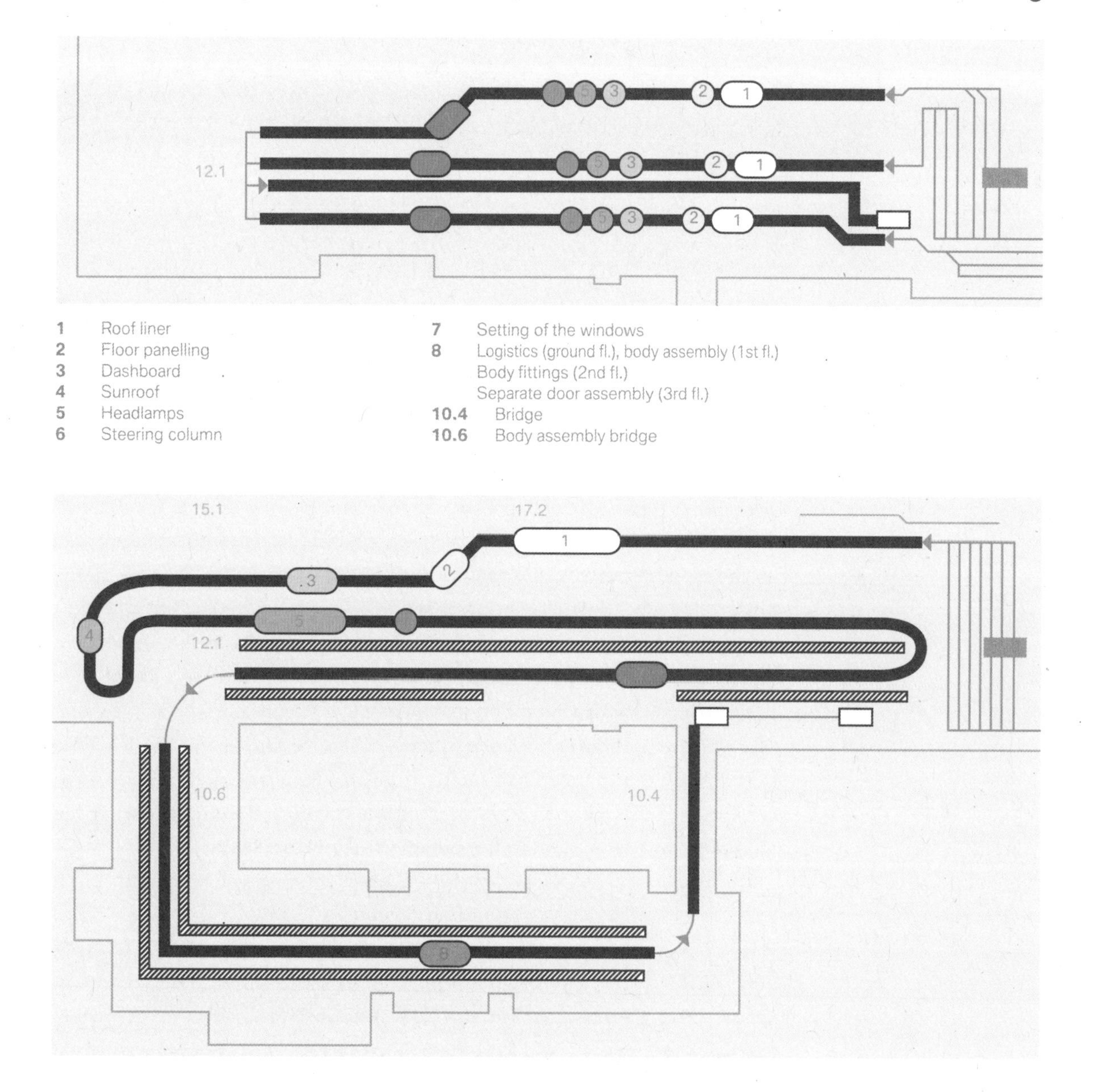

in Complex 10, which will also allow processes to be modernised; and to create a central external warehouse for outsourced parts. At the same time we also need to eliminate unattractive jobs through increased automation ... The current level of task specialisation must be reversed; job responsibilities must be strategically expanded." (*bayernmotor* 1/1983)

Aerial shot of the BMW Regensburg plant, 2002.

It was the structures on the vehicle assembly side that were in most urgent need of modernisation. Following the dismantling of Building 11, a first important step towards expansion was taken with the construction of Building 10.5. Basically, vehicle assembly was split between low-volume and higher-volume production. Low-volume production was concentrated in Hall 16 and covered everything from entire low-volume model series, such as the large-sized models code-named E3 (for example the BMW 2500), to low-volume body versions such as the two-door models of the second-generation 3 Series (E30/2) or right-hand-drive versions. Higher-volume model series or versions on the other hand, like the 02 Series or the four-door versions of the 3 Series, had their interior and powertrain installed in Hall 17. This basic split remained in place even after the restructuring of 1984. However, the integration of Hall 17 with the Complex 10 buildings via bridges 10.4 and 10.6 brought a significant increase in the total production area available for assembly of the high-volume models. The aim of the expansion was not primarily to be able to assemble more vehicles – after all, even by the mid-1970s the Munich plant was building close to 200,000 units a year. Rather, the aim was to provide the necessary space to cope with increased vehicle content in the form of more extensive standard and optional specification and additional functions.
Interlinking the assembly lines also brought other major benefits. Under the old structure, parts supply had sometimes been impeded by poor access and narrow thoroughfares. Because many tasks in the highly segmented assembly process were not logically grouped together, around half of the

parts had to be stocked at multiple locations along the assembly line, which of course took up more space. Any production defects, even if they were detected in good time, could only be rectified after final assembly, which meant that finished vehicles sometimes had to be dismantled and assembled all over again. Vehicles with the sunroof option had to be fed out from the assembly line, fitted with this option, then fed back into the line by hand. The main advantage of interlinked assembly was that it freed up a lot of space so that materials now had to be stocked at only one location and around half of the parts could be stored in small containers and flow racks. Handling equipment, like that used to install the battery, created easier working conditions by taking heavy lifting and carrying tasks away from assembly line associates. Height-adjustable assembly platforms likewise offered improved ergonomics, since the working height at the various assembly stations could now be better adapted to the needs of the associates. The transfer of bodies from one assembly line segment to the next was automated. Quality controls at the end of each assembly line section meant that less rework was postponed to the finishing stage. Some of the rework requirements could be carried out while the vehicle was still on the assembly line. The dedicated sunroof line, meanwhile, on the west side of Building 12, was now integrated in the main assembly line.

1985 saw the launch of separate door assembly, which brought numerous ergonomic improvements. One of the first vehicle assembly operations is to remove the doors, which are then pre-assembled on the string of pearls principle on a separate line, before being reunited with the vehicle in the final assembly area. This creates more space for the workers on the assembly line and protects the doors from the risk of being scratched while other work on the vehicle is being carried out.

The separate door assembly line, c.2005. Ergonomically correct working heights allow assembly operations to be performed more easily.

Transport

Many associates use buses as a relaxing and safe way of getting to and from work. This takes much of the stress out of the daily commute, even for associates who live a long way out of the city. Company-subsidised bus travel gives associates an even greater incentive to leave their vehicle at home. The company season ticket for the Munich public transport network is a further contribution by BMW to reducing traffic volumes in north Munich.

Better flow

A reorganisation of supply logistics allowed the Munich plant to significantly improve its materials flow. The new system was based on the Eching Supply Centre, which has served as an external warehouse for the Munich plant since 1985. Materials converge on Eching like threads in a spider's web. Incoming goods are temporarily stored at the Supply Centre before being made up into economical truckloads for delivery to the plant. This is a much more cost-efficient solution which at the same time also leads to a significant reduction in freight traffic. The trucks are required to follow set routes, which amongst other things prevents them from making diversions through narrow residential streets to avoid traffic hold-ups.

Engines in a category of their own

On the engines side, the company now proceeded to map out a clear strategy. Production capacity in Steyr was rapidly expanded. The Steyr plant would in future produce both diesel engines and six-cylinder petrol engines. Some six-cylinder petrol engines would also be built in Munich. Following discontinuation of the M10 generation, production of all four-cylinder engines was transferred to Steyr, although Munich continued to build the so-called special-category engines, namely high-performance engines for the M models. In the 1980s, these engines comprised four- and six-cylinder units for the M3, M5 and M635 CSi. For these units and also for the first German post-war 12-cylinder engine, a state-of-the-art special engine line was set up in Hall 140.
"Separate from the assembly lines on which [...] the 'regular' engines are built to very strict requirements, Halls 84 and 140 are devoted to production of highly 'exclusive' engines, at a present rate of 90 per day. Two-thirds

Parts and components on their way to the production lines – no more traffic is generated than absolutely necessary.

The driverless conveyor systems in special-category engine production in the mid-1980s.

of these are 12-cylinder units. The hallmark of these special-category engines [...] is a much more intensive inspection regime. [Some 150 associates] build the 12-cylinder engines, all the four-valve engines for the M3, M5 and M6 models, in any required order. Special-category engine production in Munich uses all the very latest production and assembly equipment, for example driverless conveyor systems. Inductively guided flexible assembly stands [....] travel from assembly station to assembly station, where skilled specialists process the issued materials using sophisticated and expensive special tools. This is a 'factory of the future' that's already here today, and the beauty of it is that all the computer-controlled functions and processes can be changed and adapted at any time. That means that BMW could use these same machines to build quite different engines from what it is building today. All that would need to be changed would be the computer program." (*bayernmotor* 11/1987) So Munich's engine shops was able to retain some highly attractive product lines, although the big question on the minds of the associates and works council was: would this be enough to prevent job losses? "Engine production is the heart of this plant, and that's the way it must stay," stressed the new chairman of the works council, Manfred Schoch, speaking at an associates' meeting in November 1987.

Work on the cylinder head of the M70 12-cylinder engine.

March 1992 marked another milestone in the history of engine production in Munich, when the first eight-cylinder engine in 40 years, the M60, went into production. This went some way towards allaying the concerns. To coincide with this, new production structures – e.g. combined cylinder head and engine assembly – were also introduced, along with a number of innovations in the production process, such as fracture-split connecting rods. Fracture points are built into the connecting rods during stamping, then the component is hydraulically cracked apart. During assembly, the rough fractured surfaces are then precisely remated and bolted together again.

In the early 1990s the engine production range in Munich comprised the 2- or 2.5-litre M20 (often referred to as the "small six-cylinder" engine), the large M30 six-cylinder engine (up until 1994), the six-cylinder S38 for the BMW M5, and the 12-cylinder M70. The M 30 was in production in Munich for a quarter of a century, during which time approximately 1.5 million units were built.

1992 The M60, which went into production in March, was the first eight-cylinder engine in 40 years

View of the foundry from the southwest, c. 1990.

The foundry moves out

In the 1970s, the foundry produced components for cars, motorcycles and racing cars, including alloy wheels. The 800 foundry associates produced an annual 5,200 tonnes of castings. In the early 1980s, Munich also built the first light-alloy cylinder heads for six-cylinder diesel engines, using a new low-pressure casting technique. Clearly, the only "old" thing about the foundry was the buildings it occupied. In terms of performance, it was fit and vigorous. The foundry held numerous patents and was still developing pioneering new processes right up until the mid-1980s and beyond. The last addition to the buildings was a new hall constructed in 1978, which increased the production area to 30,000 square metres, giving an annual capacity of ten million castings. But even this was not enough to keep up with the growing pace of demand.

From an environmental standpoint, the foundry had to choose between two options: demolition of the existing premises and transfer of the foundry operations elsewhere, or a rebuild on the same site. Retrofitting of emissions-control equipment was not an option. It would have been costly and would still not have achieved satisfactory emissions levels. Since a rebuild on the existing site would have made little sense in view of the increasing demand for space from the other production sectors – not to mention the need to absorb the costs of the lost production while work was under way – it was decided to transfer the foundry step by step, over a number of years, to Landshut. The Lohhof pilot foundry quickly demonstrated that much more humane working conditions could be provided for foundry staff in a new plant. In consultation with the works council, a plan was put together for the 800 associates affected by the relocation – although they were not obliged to accept the package. On 26 July 1991 another chapter came to a close and the Munich foundry finally slipped into history.

Helmut Ingrisch (2006, Works Manager 1990 – 1996)

If I had to list the five most distinctive features of the Munich plant, they would be: firstly, there isn't exactly a surfeit of space, either inside or outside the factory. This always played an important part in planning decisions. Increased efficiency could only be achieved through technology and innovations, never by physical expansion. We never had the luxury of being able to think: let's simply add more space. We were always forced to 'think smarter'. The second feature was the multicultural workforce. For management, particularly at shop-floor level, this brought a special set of challenges. Dealing with the different cultures on a day-to-day basis was a lot more demanding than what people commonly imagine when they talk about integration. Point three is the close proximity of the factory to the development departments. As the boundaries between development and production became more fluid, so the collaboration began to revert more to the way it had been before, with developers simply dropping by for a chat with their colleagues on the production line. Design for manufacturing, that is to say ensuring manufacturing issues are taken into account in the product development process, is something Munich does very well. Fourthly, the Munich plant leads something of a goldfish-bowl existence, so it always has to be on its best behaviour. It sits in the shadow of the BMW Tower and prominent guests of the board of management are regularly in and out of the plant. Fifthly and finally, Munich has always had a highly developed awareness of its own history – we have a strong sense of where we have come from, of where BMW's roots lie.

The tool shop moves out, the plant tidies up

The 1980s also saw the departure of the tool shop, which had found a new home just a few hundred yards away, on the site of the former Bundesbahnhalle, north of Moosacher Straße. The former transfer point for sheet metal and semi-finished products now became Plant 1.3. The phased relocation began in 1983 and was completed in summer 1990.

In the 1950s the tool shop handled repair jobs for the press shop and body shop. In the 1970s, its spectrum of operations was then widened to include production of press shop jig and tool equipment and welding equipment. Today, the Munich tool shop is a high-tech enterprise employing several hundred highly qualified staff.

In the mid-1980s a reorganisation and general sprucing up of the plant took place, particularly on the northern side of the site. Formerly the least prepossessing part of the plant, where all kinds of materials tended to be stored pretty much indiscriminately, this area now underwent a systematic greening. In 1984/85 a new sorting and recycling facility, serving the entire BMW Munich site, went into service in Building 47. North of Building 100, meanwhile, urgently needed parking space was created, while forklift truck

Precision work on a model for the production of pressing tools for the bonnet of a BMW 3 Series (E 21), c.1977.

The south side of Building 113. Today, this building houses the factory training centre, a factory restaurant and parts of engine production.

operation north of Building 84 was now banned. Particularly as far as this north side of the plant was concerned, local residents were closely consulted on all the plans. Meanwhile a new training and development concept, which evolved out of the former "learning centre", was implemented in the form of a new factory training centre in Building 113. A further contribution to reducing noise from the plant was made by the opening of Gate 6 for trucks in Dostlerstraße. This allowed truck traffic to be shifted to an area which did not adjoin residential properties.

Products

In styling terms, the changeover from the first to the second generation of the 3 Series was a relatively subtle affair. With the third generation (E36), however, it was time for something much more bold. Awaiting the journalists who came to the launch at the BMW Miramas test site in southern France in late October 1990 was an elongated body featuring particularly elegant lines. In addition to the radically altered silhouette, the eye was also drawn to the smooth front end, where the twin headlamps were housed behind a single headlamp cover. In another break with the past, the new 3 Series Saloon was initially offered as a four-door model. It offered occupants more space and comfort, despite only a small increase in size over its predecessor.

A BMW 3 Series Compact model (E36/5) in the finishing sector. Utmost care is taken with the final inspection to make sure that every vehicle reaches the customer in perfect condition.

For example, there was 30 millimetres more knee-room for rear-seat passengers alone. The basic shape was decidedly sporty, however, with the long wheelbase of 2,700 millimetres, short overhangs and front and rear track width of 1,418 and 1,431 millimetres promising excellent handling qualities. The sharp styling was matched by sharp chassis engineering, with spring strut suspension at the front, while at the rear, central link suspension enhanced active safety even further and at the same time brought gains in ride comfort. The engineers had invested heavily on the passive safety side, too. The new 3 Series was the first German-built car sold on the German market to feature integrated door reinforcement. In other respects, too, the passenger cell of the 3 Series was designed to the latest state of the art. The body complied with all international crash safety standards, and boasted a reversible bumper system that minimised damage in frontal collisions up to 15 km/h. In 1994 the 3 Series Compact entered the fray. 4.21 metres long, with space for four to five people and/or some luggage, and attractively priced, it had two doors and a large tailgate. Three years later, BMW

BMW 328i, 1996. The third generation of the BMW 3 Series, in production until 1998, remained much in demand throughout its life. The E36 is still a familiar sight on the roads today.

added the top-of-the-range new 323ti Compact model, which was unprecedented in this class. With its six-cylinder engine, in conjunction with sharp-handling rear-wheel drive, this sporty newcomer cut an excellent figure in its competitive class. From the outside, the main way to tell the 323ti apart from the other Compact models was by its sports suspension, which lowered the body by 15 millimetres, and the twin exhaust pipe. The 323ti had a 0 – 100 km/h time of 7.8 seconds and a top speed of 230 km/h. In 1995 BMW introduced significantly revised six-cylinder engines. The two most notable changes were an increase in displacement from 2.5 to 2.8 litres for the top engine and a changeover from a cast-iron to an aluminium crankcase. The new top model in particular boasted some impressive figures for its class: the 328i's fuel consumption of 8.5 litres of unleaded per 100 km under the European Community triple-mode test was more than a litre better than the average for vehicles in the class up to 142 kW (193 hp). At the same time, it also took torque to new levels for a naturally aspirated engine, with a peak figure of 280 Nm.

BMW Z1 assembly in Hall 159.

Low-volume production expertise and new work structures: the BMW Z1

With the BMW Z1, the team at BMW Technik GmbH demonstrated that a small team can develop innovative vehicles which are at the same time commercially viable.

It was decided to build the Z1 in the former pilot plant in Building 159, which had been used for pilot production of all BMW cars until this function had moved to the Research and Innovation Centre in 1986. The production area of 2,800 square metres was spread over three levels.

Five dream roadsters lined up in Hall 159: the production processes and work structure for the BMW Z1, like the whole concept of the vehicle, set new standards.

On many fronts, the unique design of the Z1, with its plastic, bolted-on body panels and retractable doors, probed the very limits of technical feasibility. That meant it demanded new approaches and numerous new detail solutions on the production side. Painting the plastic bodywork was just one of the challenges. The body had to be cooled with nitrogen to prevent discolouration after paint application. To prevent a build-up of static electricity, the topcoat had to be sprayed on using ionised air. In view of all this intensive investment in technology and processes, the sale price in Germany of 83,000 marks when the Z1 first came out seems by no means excessive.
The Z1, the first BMW roadster since the legendary 507, went into regular production in autumn 1988. Over its lifetime, 130 carefully chosen associates built exactly 8,000 largely hand-finished units of this model. The Z1 was the first example of an integrated project involving the Research and Innovation Centre. In its essentials, this kind of project management was the shape of things to come. Within the production force, too, a new quality of collaboration was achieved, not least thanks to the introduction of teamwork and

self-inspection. Numerous seminars and one-to-one meetings prepared the associates for a new production philosophy that said that quality cannot be created by quality controllers and end-of-process checking but rather is the responsibility of each and every associate. Self-inspection, and self-documentation of the work performed, were the new watchwords. Every associate checked the work of the previous associate, and his own, on the spot, rather than consigning responsibility to a supposedly catch-all final inspection process. This approach had a significant impact on vehicle quality, as did the introduction of teamwork. It was now largely up to the members of the relevant production team themselves as to how individual tasks were assigned. Some degree of rotation was seen as desirable and was actively encouraged. This too resulted in efficiency gains, and production times fell from the originally envisaged 5,000 minutes per vehicle to around 3,000 minutes. With cycle times of approximately 36 minutes, the individual associate had to be skilled in a wide range of tasks, consisting of numerous individual operations. For many production workers this was new territory, which for a time was reflected in an elevated associate turnover rate. It was some time before the associates gradually came to appreciate and enjoy their new freedoms.

Z1 production at the Munich plant was also the first sector to introduce the target agreement process, a system later extended throughout the company. Instead of the managers simply prescribing the objectives unilaterally, associate objectives were now jointly agreed by both manager and associates. This had the effect of increasing identification with the agreed objectives, while also increasing the obligation to comply.

The BMW Z1 Roadster; the first representative of the "Z" family, like the BMW 507 before it and like all later Z models, displayed all the typical roadster ingredients: a long bonnet, short overhangs, taut suspension and effortless engine power.

MIT fosters togetherness

Launched in 1988, the Motivation and Information Team (MIT) comprised representatives from all departments of the plant. The goal of MIT was to promote communication between different departments, to foster a sense of common purpose and to help associates understand the wider workings of the company. In 1990 an exhibition pavilion was set up, which now houses an associate shop. Later the MIT team was authorised to present and explain upcoming models to associates from all parts of the plant well ahead of the official public launch.

Special-purpose vehicles – special complexity too?

After Z1 production was discontinued, Hall 159 went over to production operations for so-called "special-purpose" vehicles. These primarily comprise public authority vehicles (mainly for use by the police) and vehicles with BMW Individual equipment.
Special-purpose vehicle production, pioneered at the Munich plant and later also introduced at the other plants, performs the seemingly impossible task of combining mass production efficiency with maximum individuality. The vehicles are fully integrated into the regular production process (including painting in special bodywork colours) – all except for a small number of assembly operations that cannot be completed within the cycle times on the

Electrically charged: Since July 2020, the BMW 3 Series Touring has been produced as a so-called mild hybrid with a particularly powerful starter generator and an additional battery. Among the first customers was the Bavarian Police.

In September 2020 Bavaria's Interior Minister Joachim Herrmann received twelve new emergency vehicles for the Bavarian Police from production board member Milan Nedeljković.

main assembly line. For these operations, for example fitting police vehicles with a roof-mounted lightbar, the vehicles are routed onto a separate line. For this approach to work, provision for the special equipment features must be engineered in right from the start of development. This integration was first adopted with the E36-generation 3 Series models (launched in 1990). This "small" BMW was larger than its predecessors, which increased its appeal for public authority customers. Another attraction was the fact that, for the first time, it was possible to have the special equipment factory-fitted. In the first full year, 1992, a total of 700 public authority vehicles were to leave the factory.
Since the fifth generation of the 3 Series (E90/E91) was launched, integration of the special-purpose vehicles into the regular production process has been perfected to the point where short delivery times can be arranged even for major orders.

In good hands

Jakob Gilliam, plant manager from 1983 to 1987, oversaw a noticeable sprucing up of the northern side of the plant, the first restructuring of vehicle assembly and the first important steps towards improved environmental practice. Hans Glas, who went on to become plant manager in Dingolfing, then took over at the helm in Munich from 1987 to 1990.
In the decade under review, both Gilliam and Helmut Ingrisch (plant manager 1990 – 1996) left a lasting imprint on the Munich plant.
Helmut Ingrisch first joined the plant in 1961. As the long-time head of plant engineering, he knew the plant inside out and could always put his finger on where changes were needed. At the supervisors' day in 1991, he asked the rhetorical question "Why do everything all at once?" – to which he immediately replied, "Because it can't wait." And it was true that Ingrisch was making heavy calls on his team: a new model start-up (the E36 3 Series, 1990) was accompanied by new work structures, the introduction of self-inspection, which had been successfully piloted with the Z1, an organisational restructuring which saw the introduction of significantly leaner

Helmut Ingrisch (second from left) handing over the reins to his successor Joachim Schulze, in 1996. The former board member for production, Professor Dr Joachim Milberg (left), looks on.

n-fair atmosphere at the Munich plant during the mmer party in 1988.

hierarchies, and the implementation of new working time arrangements (introduction of the four-day week, with weekly early/late shift rotation and a work-free day that moved on one day each week). According to Ingrisch, there could be no question of just "saving a bit here and saving a bit there". Fundamental changes in production were required if the Munich plant was to stay competitive within the wider BMW network of plants. First and foremost among the structures promoted by Ingrisch was a new management philosophy: "We must not force anything on the associates. Management and associates must jointly create the structures and freedoms that will allow the stricter new standards to be met – for example by extending associates' responsibilities or freeing them from administrative tasks. The time has come for a new way of thinking." (*bayernmotor* 5/1991)
As far as the role of management was concerned, the emphasis was now on "people management". Now, when technical issues came up at team meetings, the associates were frequently told: "You, the people who do the job, know what will work best, not us." Ingrisch also superintended the introduction of teamwork in volume production, as part of his New Assembly Structures project.
Further important priority areas for Ingrisch were environmental protection and good neighbourly relations with local residents. It was during his time as plant manager that the Neighbourhood Forum was opened and new topcoat lines were built. The link-up with the district heating network of the city of Munich in 1993, which did much to improve air quality in north Munich, was also actively promoted by Ingrisch.

Neighbourly relations and environmental protection

By the mid-1980s, a much greater awareness had been gained that environmental protection can deliver real economic benefits. One of the most powerful arguments as far as the plant managers were concerned was the savings they stood to achieve through recycling. In 1988, a new broad-based strategy for waste management logistics was introduced, while in 1990 the function of Chief Environmental Officer was hived off from facility technology and developed into a department in its own right, which took responsibility for waste, emissions and water rights for the entire Munich site. On packaging and residual materials, the order of priority was now "prevention, recycling, disposal".

One of the biggest projects the environment team were called on to deal with was the demolition of the foundry, where the old electroplating shop had caused ground contamination with heavy metals. An extensive package of measures was now required to protect both local residents and dismantling personnel, but eventually the contaminated soil could be remediated and removed from the site.

To reduce truck traffic, outbound shipments of new vehicles began to be switched from road to rail. In 1993, vehicles bound for the US, Canada and Japan left the plant by rail for the first time. One advantage of rail transport direct from the factory door is that the cars can be transported direct to the relevant shipping point (today Zeebrugge and Bremerhaven) without reloading. As far as transporting cars by road was concerned, from 1984 the trucks used were fitted with more powerful engines to reduce rpm and make vehicle movements quieter.

Gradually, a dialogue developed between the plant and the neighbourhood. This was not just confined to residents' meetings and one-to-one discussions about specific problems. For example, when a summer party for associates and their families was held to mark the "Festival of Nations" on 18 June 1983, local residents were invited along too.

Complaints from the factory's neighbours were now taken very seriously, and both sides joined forces to get to the bottom of the problem. If the plant was found to be at fault, it took steps to address the residents' concerns as quickly as possible. The following case is just one example:

In the mid-1980s, residents in Lüneburger Straße were very puzzled when something began causing vibrations in windows and glassware. The symptoms were very much like those of a minor earthquake. Initially, both the plant and the residents could only speculate as to the cause – until they got in touch with a geologist. Seismic measurements now revealed an increase in groundwater flow velocities in this area, and the problem was eventually traced to a honing machine in Hall 84. This prompted the plant to reorganise the relevant production process and return the machine to its original location. Such a problem is unlikely to recur: in 1993/94, the plant had extensive geological and hydrological surveys carried out at the site, the data from which was compiled into a digital model that was able to provide much more accurate predictions of groundwater flows. The data has since been used to calculate the likely effects of civil engineering projects on groundwater flows and to decide whether groundwater drawdowns might be required.

Since the mid-1990s, the department for environment, health, safety and ergonomics has regularly issued an environmental report on all ecologically

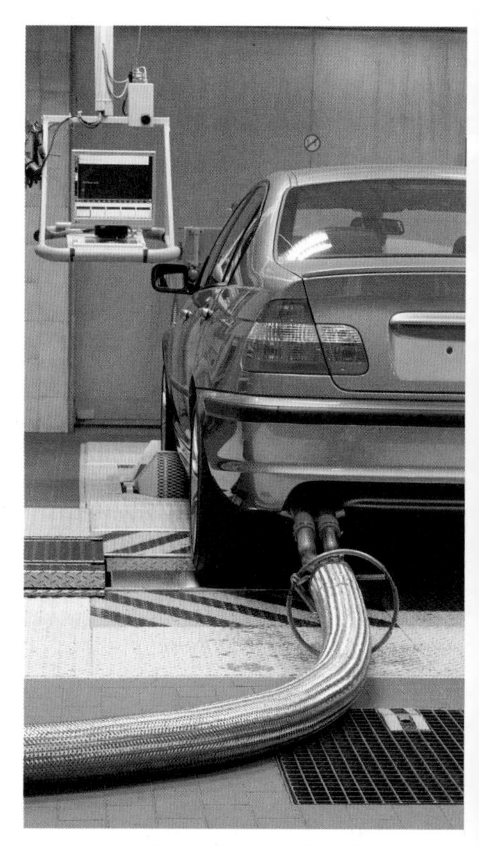

Production control (emissions testing): the central emissions testing department at the Munich plant as the emissions and fuel consumption "centre of competence" for BMW's European plants.

The Neighbourhood Forum in Riesenfeldstraße.

relevant activities at the plant. At the same time, the environmental impact of production processes has been systematically measured and continuously improved. This involved establishing an environmental management system and integrating it into the existing quality management system. In 1997, the new environmental management system was certified as compliant with the strict requirements of the EC Eco-Audit regulation and a short time later was also certified to the international ISO 14001 standard.

Extensive investment at the paint shop, meanwhile, led to a steady improvement in emissions. A new treatment facility, which went into operation in 1987, allowed grinding water and rinsing water to be recycled. In 1989 a pilot system for treating waste air from filler application was also introduced. However, no major breakthrough was possible as long as the plant continued to use conventional solvent-based technology.

In spring 1994, a new Neighbourhood Forum was opened in Pommernstraße, comprising a large lecture room and two seminar rooms: "For many people in Milbertshofen, BMW is still something of a closed book. One of the main tasks for the Neighbourhood Forum will be to rectify that,

with information events, lectures and round table discussions. ... Not least, the new Forum will also be available for use by the many community groups and initiatives in north Munich for their own events. There have already been a lot of enquiries about this, which shows just how great the need is." (*bayernmotor* 4/1994)
In November 1997, this Neighbourhood Forum then moved to a new location in Riesenfeldstraße. As can be seen from a glance at the bookings calendar, the local residents are continuing to make good use of this service. Good neighbourly relations have also been helped by the fact that for many years now, the plant has always given local residents detailed information about any major construction works or other projects likely to cause inconvenience.

Quo vadis Munich? Time for another mission review

The time was ripe for some radical measures to ensure that an industrial plant of this size, situated in the middle of a major city, could continue to enjoy the support of the community. One concern was the paint shop, which had to contend not only with environmental issues but also with capacity problems. It was the bottleneck in the plant and could only achieve the necessary throughput with three-shift operation. It was clear that sooner or later a radical modernisation would be needed that would have to go much further than the previous measures if Munich was to remain state of the art.
But first it was time to check whether further investment at this site really made sense. After all, the company had recently, at great expense, set up new production plants elsewhere that could easily be further extended. Did a company that in 1992 began building a new plant in the US, in the middle of a downturn in global car sales, still really need a plant in Munich? And was this very cautious company, whose formative experiences included a near-takeover by the Daimler-Benz AG in 1959, not perhaps in danger of building up excess capacity?
In fact, however, there were many good arguments in favour of keeping the Munich plant going. The first was cost efficiency: in many cases the Munich plant was working with buildings and production facilities that were already fully written down. What's more, cost-consciousness was

The BMW Spartanburg plant, South Carolina. More than five million vehicles have left the US plant since September 1994.

central to the team's whole modus operandi. Their careful planning and budgeting, particularly as far as human resourcing was concerned, meant that excess capacity – so often a byproduct of modernisation – was never a problem. Even during the expansion of the international production network, investment was still being carried out at the Munich plant to promote even greater efficiency and reduced costs. The Munich team seemed to thrive on the motto "The only constant is change" – not only in terms of their willingness to ride the 1980s trend of major structural change, or their flexible attitude to new technologies, but also in their willingness to embrace new ideas and new ways of working at the ordinary day-to-day level, and to take an early, proactive role in shaping structural changes rather than simply sitting back and accepting change passively. It also helped that the works council adopted a critical but always constructive attitude to the necessary adjustments of the 1980s. And last but not least, the continuing strong growth in demand for BMW cars made Munich's production capacity indispensable.

Electrostatic painting (the electrically charged paint is attracted to and deposited on the grounded body). / Bodies after passing through the final inspection in the paint shop.

Two steps forward: new assembly structures and new topcoat lines

In 1993 the first stage of dismantling work at the former foundry – the second stage followed in 1999 – created space for two new topcoat lines. A topcoat line is where the colour of the customer's choice and the protective clearcoat are applied to the bodywork. Construction of the new topcoat lines began in April 1994, and by January 1996 they had already gone into service. The new topcoat lines also introduced water-based paint technology at the Munich plant, an important advance in environment-friendly production.

With the launch of the New Assembly Structures project, the mid-1990s saw an even more radical watershed than ten years before. This reorganisation introduced a philosophy that has been a hallmark of BMW ever since – the single-line assembly principle, whereby all models and versions take shape on a single assembly line. The assembly process, together with the painting process, is the main conduit for offering customers individuality and diversity. Therefore the assembly process must be geared to offering the widest possible choice of specification. The many possible permutations of optional equipment features and engines, and the need to comply with different national legal requirements, inevitably results in a high diversity of parts. The single-line principle supports this with efficient logistics and assembly processes.

The limitations of the old structures had become particularly clear with the E36 3 Series, which went into production in 1990. It was therefore decided that production of the successor to the E36 (the E46-generation 3 Series, which went into production in 1998) should dispense with these problems. Not only that, future production lines would deliberately be designed with no provision for taking vehicles out of the line during assembly. The various line sections would form a single continuous main line (interlinked assembly line). This meant that only small amounts of rework could now be performed in-line, which introduced a salutary pressure to get things right first time. In the case of new product start-ups, this makes heavy demands

Robert Hohmann (Assembly)

I started my training as an engine fitter in September 1976, and by 1979 was already working as a maintenance engineer on the engine production line. Those were good times. I enjoyed working on hydraulic systems, pneumatic systems and transfer lines. It was very formative, and from that time onwards I've always enjoyed my work. After training as a foreman from 1986 to 1989, I was given a foreman's post right away, in wiring harness manufacturing. At that time we were also producing wiring harnesses for the Dingolfing plant. Later I was involved in planning the move to Building 46. This was a great project for me as a 'humble' foreman to be involved in. Today our wiring harnesses are sourced externally.

At one point – working on the Z8 – I was responsible for the incoming goods area and for the entire first floor, from the fuel filler neck to window fitting. It was an amazing experience: the painted body was unloaded one day and the next day the vehicle was finished. Z8 production had a building to itself, consisting of three floors. The cars were simply pushed by hand from one station to the next. We were a great team and everybody was really passionate about the car.

Then, as my 25-year service anniversary at BMW and at the Munich plant approached, I decided it was time for a change. I heard that Rolls-Royce was looking for logistics engineers for the Goodwood project, and I applied right away. I started in April 2002. That first year was very hard. We were developing everything right from scratch. But the second year was a lot better. All in all it was a great experience, despite the language difficulties at the beginning. The keys to the first vehicle were handed over at midnight on 1 January 2003. We were working on that car almost right until the last minute on New Year's Eve. Ten years ago I came back here to Munich. Although I could have moved into a headquarters post, I wanted to be working close to the product and to go back to being a foreman. I'm now a group leader at Body Assembly. BMW is a unique company. In the almost 40 years I've been here, nothing has ever happened that I thought was out of character, that I thought wasn't what you'd expect of BMW. I virtually grew up with BMW. My father worked at the company for 40 years and my daughter's already been here for more than ten years.

Windscreen fitting of a BMW 3 Series in 2007: two robots insert the front and rear windscreen simultaneously (BMW 3 Series Saloon). (left above)

The tilting assembly system, in the form used since 1996. (left below)

Interlinked body assembly line. (below)

on the maturity of the new vehicle. If current-model vehicles are still being built on the same line, the whole production process will inevitably grind to a halt if the new model is still experiencing any major problems. Later, a slight concession was made and an exit point was provided after all – but only on the proviso that it was not to be used for current models. The permitted time frame for introducing the new structures was just six weeks.
For the production force, the new structures brought many ergonomic benefits. From the start of the assembly process right through to the finishing sector, the area the associates stand on now moves along in sync with the vehicle. Previously the associates sometimes had to walk alongside the vehicle on a stationary floor as they performed their jobs.
The tilting assembly system – already successfully piloted at the Regensburg plant – was now also introduced in Munich. To accommodate it, an additional level was added in Hall 16. When working on the vehicle underside, the tilting assembly system allows bodies to be tilted in a hanger when fitting underfloor parts, thereby avoiding the need for overhead work. Teamwork, successfully piloted on the Z1, was now also introduced for high-volume production. At the same time, all assembly platforms were now designed to be height-adjustable. Attached platforms allowed the associates to walk round the vehicle without having to step up or down. Following its successful debut in Munich, the single-line system pioneered by the New Assembly Structures project was later adopted as the standard BMW vehicle assembly procedure, and has therefore influenced the design of all the newer vehicle plants.

Ab hier
Arbeitsbereich

The new Munich plant takes shape

In the mid-1990s – on the back of an unparalleled programme of investment – BMW set about building a virtually new plant on this site with such a rich tradition. Production start-ups for two vehicles distinguished by outstanding quality were the deserved reward for all the effort and endeavour. New machinery and processes, countless manufacturing innovations and highly motivated associates ensured a remarkable turnaround and a spectacular renaissance for the Munich plant.

1995 – 2005

The start-up of the fourth-generation BMW 3 Series – a test

It did not take long for the group work to have an effect, as a young associate was soon to discover during a period as a "line partner" in the assembly halls. "Line partners" are white-collar associates at the company who take the opportunity to spend a week working in production. It is an experience which certainly leaves its mark.

"I gradually got to know everybody in the group. I got the feeling that they worked well together. '12 months ago,' my TM-42 colleagues told me during our breakfast break, 'there was a very different atmosphere around the place.' However, following the introduction of group work, the working structures and the atmosphere in the group improved markedly. ... 'We talk about production over the past few days, audit results, working processes, shift handover problems and anything else that affects us,' explain my workmates for the day. 'If it becomes clear during group discussions that errors are being made repeatedly at a certain point on the assembly line, the origin of the error is investigated. Measures are then introduced immediately in order to put things right. It often used to take a long time to track down errors,' adds Gerhard Hoidn, a foreman in TM-42. Every member of the group, all of whom have also been involved in discussions on set targets for a year now, has a serious interest in ensuring that quality is right. And, as they all recognise, only by working as a group can they achieve the best results. This is also something I experience – here the motto 'Only together are we strong' truly is a reality. Anytime I can't keep up – like if a door arrives quicker than expected – help is immediately at hand. According to Hoidn, as well as the feeling of responsibility, the community spirit has also grown strongly. And this is how the group can influence the organisation of work at the assembly line." (Report by Sybille-Beate Waegner, *BMW Zeitung* 12/1997)

Group work in the assembly of the 2005 BMW 3 Series E90.

TalentFactory – Apprenticeships with that 'start-up' feeling

The BMW Group is creating diverse and attractive employment prospects for young people through its extensive range of apprenticeships. With its global vocational training programme, the BMW Group is ensuring that the constant demand for young talent is met and is promoting skill development within the company. Young people are preparing for their future fields of work in 30 apprenticeships and 18 dual-study programmes at 19 locations in the production network and at the headquarters.

Innovative ideas

Vocational training at the BMW Group plant in Munich has developed and established the 'TalentFactory' as a learning method for apprentices to further promote practical, independent learning and personal growth as well as collaboration and creativity in finding solutions.

Here, apprentices and dual-study students in their first year of training work following a start-up model, where they develop products and services for internal specialist divisions and external customers. At the TalentFactory, the trainees work together in interdisciplinary teams, learn agile work methods and, at an early stage, assume responsibility for an entire project.

A passion for learning

Effectively, learning covers various processes and is always 'on the job'. The junior employees put together a suitable team for the project and also involve apprentices from other professional areas in the process of developing and implementing ideas. The trainer acts as a coach, and an employee from a specialist division follows the process as a mentor. The trainees continue to receive their theoretical knowledge at trade school and in in-house courses.

The vocational training department at the Munich plant announced back in autumn 1997 that group and teamwork now formed part of their training courses. Some vocational trainers had to update their methods in order to meet the human resources-related and specialist requirements of the new working structures.
Joachim Schulze, plant manager from 1996–2004, introduced a new target agreement process across the board, setting binding goals which each manager agrees with his or her team instead of imposing them, as had previously been the case. The process demanded a shift in thinking. "You can't just impose a process like this, it has to become an integral part of the way people actually work – and that also applies to group work. Otherwise, we'd just be sticking a new label on an old way of doing things. It is particularly important that associates within the plant system work with, rather than against, one another. It is all a question of sharing experiences and expertise. The objective is to make quality, supply reliability and efficiency trademarks of the Munich plant."

Eva Karrer
Electronics technician for automation technology (trainee)

I first got to know the BMW Group through a school internship, which is why I decided to pursue an apprenticeship as an electronics technician for automation technology after graduating. I especially like the variety of assignments in the different production divisions, where I'm constantly experiencing new things and learning in a friendly and collaborative environment. So, there are always opportunities to innovate, engage and apply what you've learnt. For example, I've already been able to actively participate in interviews, various events, and in supervising interns. I also appreciate how respectfully and collegially the trainers and trainees interact with each other.
There will be lots of opportunities to pursue a career with the BMW Group after completing my training. After my apprenticeship, I'd like to work in Industry 4.0, as the range of work in that area is very broad, varied and challenging.

Associates on the finishing line, 2006.

Evidence that the only way to achieve success was to maintain constructive cooperation with partners in the development department and the other plants arrived in the shape of Munich's next major challenge: the production start-up of the E46/4 3 Series Saloon – initially at the Munich and Regensburg plants concurrently – in 1997. A parallel start-up at two plants demands precise coordination and meticulous planning, as all those involved in development and production were acutely aware. A policy of close cooperation and an ability to think as a partnership, as well as the early integration of all the development and production associates involved, rapidly bore fruit. The plants were able to highlight to the development managers where problems lay at an early stage. As a result the E46/4 had already reached a previously inconceivable level of quality by the time it went into series production.
The first pre-series models had already been built in normal series conditions in June 1997. Series production in the paint shop and assembly got under way in December 1997, with the bodyshells initially supplied by the Regensburg plant.
Fundamental alterations to pre-series production and supply logistics strategy helped to meet the high standards required by the start of series production. The pre-series cars were divided into groups, and parts were allocated almost exclusively using series systems. A very close eye was kept on all areas of the process chain, from the suppliers to the installation point on the series assembly line. Product and process qualification was carried out with some 600 functional and pre-series cars. Needless to say,

interruptions to series production were to be avoided. The associates were put through intensive training courses in training areas set up specially in the start-up centre in Building 159 (special-purpose vehicle construction having been relocated to Hall 113).
The Munich plant had learnt a few lessons about its in-house processes. In particular, cooperation between the individual areas of production – known as "technologies" – was boosted significantly "on the shop floor": "The finishing teams are comprised of 150 associates from the quality management department, body shop, paint shop and assembly. In a combined final inspection stage, the cars are given the green light as 'ready for customers'. Here, the teams assess the cars from a customer viewpoint and confirm the pre-set delivery quality." (*BMW Zeitung* 12/1997)
All the technologies were therefore responsible for quality standards in every area of production and reworking. From now on, each technology would have to do its own reworking. This was another area where the com-

Electrical overhead conveyors made their automotive industry debut in the Munich body shop – well before the start of production of the BMW E46 3 Series (pictured here).

pany did not have to wait long to see the results. Delivery quality and punctuality – in meeting the delivery date promised to the customer – improved, and reworking costs were substantially reduced.
The start-up for the E46/4 3 Series Saloon has gone down as one of the best start-ups in the company's history. However, the 3 Series Compact (E46/5) then raised the bar once again in the spring of 2001. The experience gained from these two successes flowed into the first ever digital start-up – for the facelifted Saloon – in autumn 2001. The switchover to the updated model between the end of one shift and the start of the next was seamless.
The fourth-generation 3 Series was also a milestone for Munich's partner plant in Rosslyn, South Africa. Following intensive preparations for the start-up, in which associates from Munich and Regensburg also played a part, Rosslyn was integrated into the plant network as a fully equipped facility.
Demand for the new car, or more accurately the interest of customers ahead of its market launch, once again exceeded all expectations. Production at Munich continued uninterrupted through the summer of 1997 and 1998. Indeed, the lines even kept going through lunchtimes, and special Saturday shifts were also brought on stream.

1997 Production start-up of the E46/4 BMW 3 Series Saloon at the Munich and Regensburg plants

The BMW 3 Series Compact (E46/5) in the body finishing area.

1.2 billion euros for a "new" plant

The start-up for the new 3 Series delivered another significant boost to the home plant's prestige. The finances for the development and ongoing maintenance of the newer production locations were often signed off as all part of the replacement process for the home plant, which – it was implied – no longer had a long-term role to play. The same reasons were given for the relatively meagre investment in the Munich plant into the second half of the 1990s. The sums involved were intended merely for the upkeep of machines and materials, so that production to the desired quality standards was possible for some more time.
However, with the success of the E46 start-up under its belt the Munich team could now emphatically defend its place in the production network. It was only logical for the company to look again at its plans to phase out its

Fitting the pistons in the eight-cylinder N62, 2003.

operations at the Munich plant in 1999/2000. The plant was duly given another chance to carve out a long-term role for itself. The major achievements of the Munich associates over this period had been to keep their chins up despite the uncertainty surrounding their future, and in some respects (notably its new assembly structures) even to remain a step ahead of the game at the same time. And yet there was a lot of ground to be made up. Recent years had seen a de facto block on investment, which was visible in many areas of the plant beyond just its outside appearance. This lack of funds was threatening the very fabric of the plant. While perhaps sustainable over a period of a few years – indeed, the economic plight of the plant could have served to highlight its frugal running capacity – sooner rather than later Munich would have reached crunch point. Low investment, and the consequent lack of the necessary machinery, technology and processes, cuts off the route towards future savings potential.
In order to carve out a future role for itself beyond the expiry of the company's phase-out option, the BMW home plant had to make its case for substantial investment. It started with engine production.
The works council had been campaigning to secure the long-term future of engine production in Munich for a number of years. In the early 1990s, a new eight-cylinder petrol unit was added to the production roster and, in 1998, a decision by the board of management finally delivered the certainty they were looking for. From now on all the Group's V engines were to be built in Munich. Production of high-performance engines also remained in Munich, and the six-cylinder petrol unit – which was now a single series following the phasing out of the "large" six-cylinder variant – would continue to be built in Munich and Steyr as a volume engine.
The implementation of this decision demanded considerable alterations to the structure of the engine shop. Crankshaft manufacture and other areas of machining for the new eight-cylinder (new generation V8 petrol, N62) were housed in the new Building 144, which was connected to Hall 140 by a supply bridge. Here, state-of-the-art machining centres were installed to handle the work. In order to allow series engine assembly to be accommodated on the first floor as well, high performance assembly was relocated to the ground floor. This represented a monumental logistical challenge, since a suspension of production was not an option for capacity reasons. The move had to take place predominantly at weekends. Prototype construction and pilot assembly were moved to Building 100 to clear extra space in Building 140.
A core element of the new structure was the integreated engine storage facility (in the new Building 85). Here the engines produced in Munich are kept in temporary storage before being dispatched, usually in the order in

The restructuring brought with it state-of-the-art, pleasantly designed workplaces in Hall 140. Pictured is the "complete engine" assembly line in V-mirrored form (taken c. 2003).

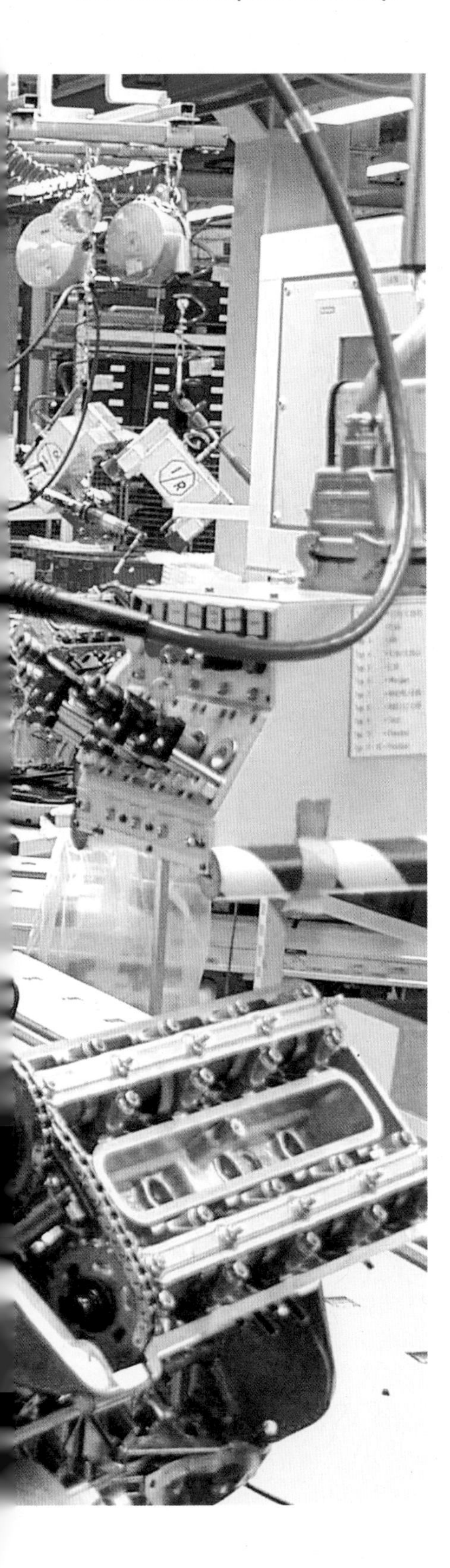

Building 100 today houses the energy control centre. Up to recently it was home of the engine production's pilot assembly.

which the destination plant requests them. The integrated engine storage facility also contains the units from Steyr (four- and six-cylinder diesels, six-cylinder M petrol engine) before they are sent off to the Munich car plant for 3 Series production. The new topcoat painting lines improved the plant's environmental record and freed up capacity when they were introduced in 1996. However, these lines were only an interim measure, with the real restructuring stage still to come. Significant improvements in quality, a breakthrough in environmental performance and notable advances in efficiency were only possible with a new pre-treatment area. The demolition of the remaining sections of the former foundry in 1999 cleared the way for the new buildings required – and a new paint shop was duly constructed from the ground up by 2003. Once again, the building work did not necessitate any extended suspensions in production.
Celebrating a world premiere in the new Building 19.5 was the rotation dip bath (completed in 2001), which could now for the first time be used right through from final body cleaning to the cathodic dip bath (CDB) stage.
Following a process of thorough cleaning and treatment with a layer of zinc phosphate, complete bodyshells are fully immersed in the CDB. With the bodyshell consisting of various different metal surfaces, the layer of phosphate is added to ensure a consistent top surface. This, in turn, serves as the basis for an even CDB coat over the outer skin and all cavities. The CDB layer forms the basis
Following a process of thorough cleaning and treatment with a layer of zinc phosphate, complete bodyshells are fully immersed in the CDB. With the

Building 19.5 (cathodic dip bath) and 19.6 (protective undercoat, sealing).

bodyshell consisting of various different metal surfaces, the layer of phosphate is added to ensure a consistent top surface. This, in turn, serves as the basis for an even CDB coat over the outer skin and all cavities. The CDB layer forms the basis for all subsequent stages of painting. At the same time, its chemical composition makes this the crucial layer in ensuring corrosion protection.

The CDB process was, as such, nothing new, having been used by BMW for many years already. An outstanding feature of the new facility, however, was the newly developed conveyor system with RoDip3 immersion process. In contrast to previous processes, here the body is not only dipped but also rotated on its axis several times inside the 11 dip baths (pre-treatment and CDB). The conveyor system is no longer located above the baths, but to the side outside of the coating area. This conveyor technology prevents dust from the conveyor system getting into the baths.

A subsequent expansion stage saw the introduction of a new protective undercoat machine and two fully automatic filler lines. The filler slots into the paint system between the CDB primer and the topcoat. In addition to providing an extra layer of corrosion and stone-chip protection, the filler tone also enhances the colour of the topcoat.

The exclusive use of robots to paint the cars allows all areas of the body to be reached even more effectively. To start with – before entering the system – the cars are cleaned fully automatically with blowers. Here, an air current is used to release tiny dust particles from the body, and these are then immediately vacuumed up. Six robots per filler line paint the outside of the

Dip bath.

Fully automatic application of the protective undercoat.

body. Rotary-type spray technology (the paint is sprayed on at approximately 45,000 rpm) and electrostatic paint charging (with approximately 80,000 volts) help to create an even application of paint over the grounded body. Another group of six robots automatically open doors and lids and paint the inner areas of the body. The introduction of the new filler systems represented the final stage in the conversion to environment-friendly water-based paints at the Munich plant. Another contributor to the environmental standard of the new paint shop is the regenerative afterburning of waste gases, including heat recovery for the paint drier, a virtually closed water circuit in the painting lines and advanced waste water treatment.
The much longed-for environmental improvements were at last a reality and Munich locals living nearby have since been among the key beneficiaries. The city of Munich presented the plant with its environmental award for "exemplary environmental achievements and work for the community" in 2003, and the Munich plant was granted membership of the "Bavarian Environmental Pact for environmentally compatible economical growth" several years ago for committing to ensuring a voluntary, qualified improvement in environmental performance.
Of the existing buildings, only Building 19.0 and Building 20.1 – home to topcoat finishing – survived intact.

Application of filler onto the BMW 3 Series E91, 2006.

2003 The Munich plant is presented with the city's environmental award

Doing things differently: KOVP

The aim of the KOVP (customer-oriented sales and production process) is to shorten delivery times, keep the option open as long as possible for customers to make changes to their order (up to just a few days before built-to-order production starts in assembly) and, at the same time, optimise manufacturing processes across the board. Each order is now assigned to the respective customers at the assembly stage in the production halls, rather than earlier on in the body shop as used to be the case.
The customer's vehicle order is fed into the production network online and assigned to a particular plant. The order is received by the assembly team, as they represent the logistical core of vehicle production. Assembly then requests a body variant in the desired exterior colour from the central body sorting system and stamps on the chassis number. Only now is the assignment of body to order finalised. The remaining stages of assembly are now carried out, including the installation of the drive train (also known as the "marriage" with the body), before the vehicle rolls off the assembly line and embarks on a series of exhaustive tests. KOVP therefore turns the early order assignment system of the 1970s, as described in Chapter 8, on its head. The current process is the only way of ensuring (with the help of a central storage facility containing standard-trim bodies) that customers can change their mind on the colour of their car as little as a week before the start of assembly. The order is finalised just a few days before the start of assembly, and the suppliers (including engine production) are informed which parts are required at the assembly line in what order and at what time. The production team then enjoy the benefits of significantly smoother logistics, the ability to paint bodies "in convoy" (up to 20, one after the other, in the same colour) and considerably improved storage logistics between the paint shop and assembly. The bodies are "pre-produced" as far as the paint shop stage, where assembly then switches to an exclusively built-to-order footing.
The new assembly structures and the IMOLA engine storage facility were already prepared for the KOVP model, which had been introduced gradually. Now it was just a question of getting assembly totally ready for the new process – and that meant a central body sorting system was needed, stocked with bodies which were as identical as possible (except for their exterior colour). In 2001 a second storage facility was set up between the paint shop and assembly in order to provide the required capacity.
Thanks to KOVP, the apparent paradox of efficient production processes on the one hand and the virtually unlimited scope for customers to

The body sorting system in Building 19.4. Here painted bodies await the call from the assembly department.

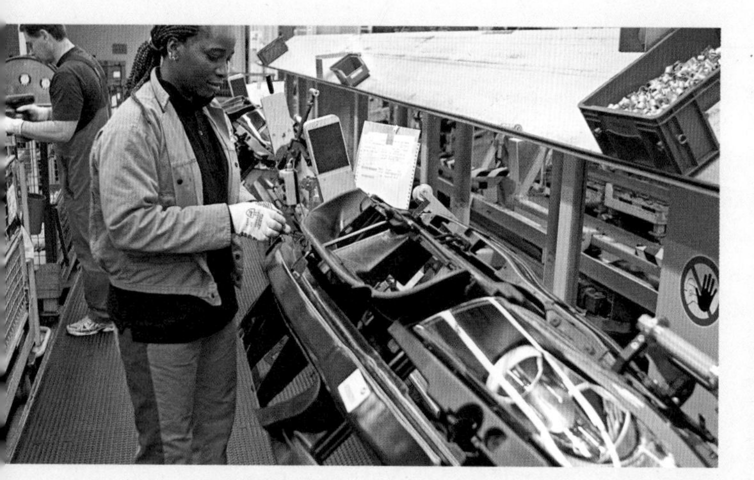

Despite extensive demolition work – the former Building 23.0 had to give way as part of the plant's development – the building regulations stipulated that no dust should be created by the construction site for the new Building 16.1. To this end, the site was screened off to impressive effect. For example, a partition wall was built specially to protect the neighbouring Building 16.0. Construction work on enclosing an existing, still-in-use conveyor bridge within a new bridge had to continue at weekends, as this was not an area where the work could be carried out while production was in progress. An extremely considerate construction process – known as "Berliner Verbau" (Berlin-type pit lining) – was used for the foundation work to largely eliminate vibrations and spare adjacent production areas and people living near the plant as much disruption as possible. It was thanks not least to this thoughtful approach that no complaints about noise or dust were received from neighbouring residents over the entire course of the construction project. Care was also taken to protect the environment while the work was taking place. Unusable material from the demolished building was examined, classified according to its pollutant hazard and disposed of without requiring long-distance transportation. Leftover concrete from the build-ing could be used again in road construction, while dismantled steel girders were recyclable.

individualise their vehicle on the other has therefore been largely resolved. In order to ensure that this process not only runs efficiently but also that errors are kept to a minimum, the more complex modules are pre-assembled ahead of the main assembly line. In this way, scope for errors on the main line are further reduced.

This approach spawned the addition of another clutch of pre-assembly areas at the Munich plant, offering two further benefits: they allow improved space management – a valuable asset in Munich in particular – and the cycle times in force on the main line do not necessarily need to apply in the pre-assembly areas as well.

In autumn 2004, shortly before the start of series production for the latest 3 Series, four models from two generations (E46 and E90 3 Series Saloon, E46/5 3 Series Compact, E91 3 Series Touring) were assembled at the same time – thanks in part to the variant-neutral main assembly line. Minor adjustments were also required in the body shop for KOVP, but the main requirement – for a "standard body" – had already largely been met with the fourth-generation 3 Series. For example, there were now only eight variants of the Saloon, compared with over a hundred for the previous-generation model. In 2003 work began on a largely new body shop, which was intended primarily for the upcoming new model.

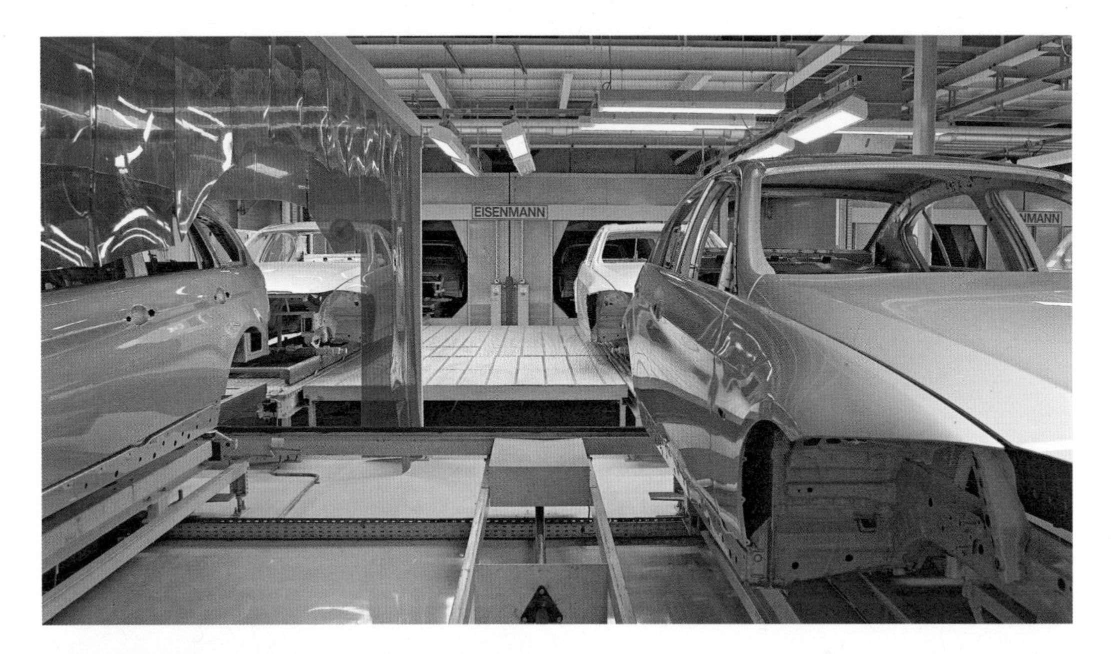

The end of the drying process [Paint Shop] (above) and painted bodies, ready for a final checking of surface quality. (below)

Side frame production and the storage level in the Body Shop, 2006.

Enter a team of new robots ...

The fifth-generation 3 Series was built as of late 2004, using the largely identical machinery in the Munich, Regensburg, Leipzig and Rosslyn plants. To make this possible, far-reaching harmonisation measures were required at the plants, in particular in the underbody production and body framework sections. To this end, the Munich body shop constructed a new building (no. 36) with a link to Buildings 32 and 33, which were already used for body production. Plus, a new side frame production area and machines for the production of components for the assembly section were also added. The restructuring meant that, after almost a quarter of a century, body construction could be brought back together into a single complex.
The new body framework system, breaking new ground with its impressive complexity packed into the tightest of spaces, represented a technical tour de force. A team of 12 robots, working to 56-second cycles alongside each other like cogs in a perfectly oiled machine, occupied a surface area measuring only 820 each with eight variants, were fixed together with some 90 weld spots. This system allowed two models to be produced without time being lost in each cycle through switches between machinery. In order

The body framework system, 2006. The robot on the left brought the outer section of the left-hand-side frame into position, while its partner pictured right prepared the roof for assembly onto the body. The body at the top of the picture is on its way into the geometry cell.

The machinery (punch and moulds) used to produce body parts, 2005.

The very beginning of automotive production: stage one sees rolls of steel (coils) being rolled out and cut into flat sheets, or plating. The plating is then re-formed into body parts.

to ensure that the system also worked in series production, the 12 geometry cell robots were first put through their paces virtually in a 3D simulation program at the Research and Innovation Centre in Munich. Sharing its construction with comparable units at other BMW Group plants, the system in Munich was the most efficient.

… that need regular feeding

Further increases in productivity were on the agenda in the press shop. Between 1998 and 2007 the number of re-formed parts almost doubled. The press shop introduced a series of measures aimed at reducing vibration levels, including fitting some of its pressing machines with steel-spring damper mounts.

As part of the BMW production network, Munich used to produce body parts for the Dingolfing, Regensburg and Leipzig factories – and was also supplied with body parts by Regensburg and Dingolfing.

Two pressing lines in Building 154.

Quality is what others deem excellent

The Munich team's remarkable efforts and the impressive programme of investment were to reap rewards when the wider industry backed up what the plant's internal quality statistics had long been claiming. On five occasions between 1998 and 2005 the Munich plant was ranked in the top three European car plants in the Initial Quality Study carried out by renowned US institute J.D. Power & Associates. Indeed, in 2001 and 2003 it earned the title of "Best plant in Europe" in the study.

In its annual car plant quality assessment report, J.D. Power investigates the number of problems which owners experience in the first three months after purchasing their new vehicle. The study, which represents an important quality gauge for the automotive industry, takes a sample of 100 current model year vehicles and looks at over 100 possible problem zones in nine different areas.

New ways of working, such as group work, and countless quality-enhancing measures were now paying dividends. And there were more innovations still to come. Chassis construction was one example. Introduced in 2004, the referential axle geometry testing system (RAMA) allowed the chassis settings with the most significant influence over straightline stability, steering precision and directional stability, in particular, to be measured down to an accuracy of 1/100 mm. Where previously measurements could only be taken at certain points along the stationary vehicle, continuous measuring was now possible. Handling characteristics could be recorded on the test rig with all suspension settings, allowing the engineers to collect more meaningful data. The rig was the product of a jointly-run project with the development department. It was also used in the Research and Innovation Centre, its greater measuring precision focusing first and foremost on identifying at an early stage possible areas for improvement in the development and start-up phase. At the Munich plant the rig's role was to safeguard series production quality in random vehicle audits. The axle measuring rigs in the assembly checking zone, which all vehicles were sent through, worked with the same technology.

RAMA in action. In-depth testing has been carried out on vehicles in Building 163 since 1981, with the aim of safeguarding production quality. The initial check lists already contained as many as 900 items.

Safely does it

In addition to a raft of other certifications, since 2004 the plant has also been certified for the application of a management system for work and machine

Test rigs for adjusting the track and camber of a vehicle's front axle and aligning its headlights (2006).

It's a girl thing

In 2002 the Munich plant hosted its first ever Girls' Day. Every year since then schoolgirls have had the chance to spend a day in a technical trade. The percentage of female trainees on technical programmes has been on the rise for several years now. Many other initiatives are also helping to attract increasing numbers of schoolgirls from various partner schools to vocational training in technical trades.

Mehmet Özsoy
(2006, Body Shop)

I have worked in the body shop at BMW since 5 November 1979. My colleagues welcomed me into the team immediately. When I started here, I was working eight hours a day. That meant walking eight kilometres every day along the line and lifting a total of four tonnes in weight a day. There were no robots back then. I had to lift each door for the 3 Series Convertible two or three times, and they weighed 18 kilos each. A lot has changed now, though. Today the guy who does what used to be my job no longer has to lift heavy weights. Later I worked as a quality specialist. If there were problems with a door, for example, I often used to go back into work to help sort them out. Then I became an assistant foreman. Now I'm a methods expert. I'm very proud to work for BMW.

safety in accordance with the OHRIS (Occupational Health and Risk Management system) guidelines produced by the Bavarian state authorities.
Given the status of the plant as an almost permanent construction site during the period examined in this chapter, as well as the space restrictions at the plant, the topic of work safety carried extra weight. Today, the department responsible for work safety, ergonomics and environmental protection exerts an influence on the structure of workplaces at a very early stage of the planning process. The work safety team has worked hard with the various specialist sections to reduce and remove danger hot spots. Statistics have shown the number of accidents to have fallen sharply over many years, and the principle of "work safety as a responsibility of management" has reaped impressive rewards. The "Work Safety Cup" competition has taken place at the plant every year since 1988 to find the safest departments, and this was joined a few years ago by the "Continuity Cup" in recognition of several years (i.e. a sustained period) without accidents.
There's no room for excuses. Even at a plant where some of the buildings are getting on in years the associates have a right to expect ergonomically designed workplaces. The latest example of this is the use of height-adjustable assembly platforms in the wiring harness fitting section, despite its extremely low floor-to-floor height. In six work cycles, which see the associates inserting and routing wiring harnesses, screwing in control units and putting together plug connections, they no longer have to climb into

the vehicle. Instead, the height adjustment facility allows them to perform a large proportion of their work standing up and with an ergonomically favourable body position. Plus, wall bars have been installed in the immediate vicinity of the workplace, which associates can use to do exercises aimed at relieving the stress on the spinal column.

2004 On 7 May 2004 the 7,000,000th vehicle rolled off the assembly line.

Excellent occupational safety: the winner of the 2019 Occupational Safety Award, Klaus Murr, with plant manager Robert Engelhorn. (top) / Ergonomic workplace: Employee installing the tank on the 3-series plug-in hybrid (2018). (bottom)

Fraukje März (Head of Quality Management)

The exciting thing about production is when a new series launch is due and these vehicles are gradually integrated into the existing production system. The many cogs start slowly turning until they run perfectly together. When the first cars leave the plant uncamouflaged after the end of the confidentiality period, it's a great feeling. I know I'm part of it! And so, we complete one smooth launch after the other, and have actually built up a great reputation in the production network. We take our role in this network very seriously and support our partner plants in Mexico and China.
Our team knows the meaning of working in the midst of a city and to be considerate of our neighbours. We also know that we need to be efficient to remain competitive. It's good that we plan our finances to be tight. It promotes creativity. It's how we did projects in the past and how it is now when we plan new structures for the body shop and assembly. I'm convinced that we'll manage it well again this time.
Before I started my job as the foreman in body assembly, I thought BMW was like a big train, where the individual doesn't really have any say in the direction of travel. Not at all! It's entirely possible to make changes. Being foreman was a really special experience. I learnt a lot about how a production system works and I look back affectionately on that time. I'm now in charge of the plant's quality management division. It's amazing how our employees and managers identify with the company. It's a precious thing for BMW employees to have such loyalty, especially in such a large company.

Have you ever seen anything like it? An almost parallel start-up in four plants

The start-up for the fifth-generation 3 Series highlighted the flexibility of the plant's working hours to particularly striking effect. The team was able to perform many assembly changeovers without suspending production. However, the switches in the assembly section – affecting the areas described as well as virtually every other cycle – and in the body shop required suspensions in production totalling eight weeks during a period of just over a year. Flexible working times were a major factor in allowing the plant to avoid having either to release associates or to accrue overtime costs. Associates were able to take several weeks of summer leave as usual in 2004 and 2005, something those with roots in Turkey, Greece and the former Yugoslavia appreciated in particular.

On 1 December 2004 the new BMW 3 Series Saloon, the E90, went into series production at the Munich plant – the launch factory for the new BMW 3 Series. This meant that Munich was also preparing the ground for the start-ups at the other three E90 plants in Regensburg, Leipzig and Rosslyn, where production of the new model got under way as early as February and March 2005. Associates at the Munich plant were involved in building the initial prototypes at the pilot plant in the Research and Innovation Centre more than a year before the start date for series production. Experiences at other plants, especially the excellent start-up for the 1 Series in Regensburg, were analysed thoroughly in Munich. Here, the Regensburg team

Topcoat finishing.

Assembly: building on the front end.

had contributed some fine preparatory work for the use of the production network as a whole. As the largest section in production, assembly had to carry out the most extensive training courses. All associates spent a period of one to three days, depending on the content of their work, in specially built training stations, where they were given preparation for their new jobs. As part of its leading role, the Munich assembly team took responsibility both for process development and structure at the plant and, above all, for product qualification and the technical assembly of the new 3 Series. It was a task made all the more challenging by the varying structural conditions of the four locations. Never before in the start-up of a new model had so many vehicles been built to such high quality standards in such a short space of time. Demand was huge from the word go, with the Munich plant producing 13,000 showroom models for the BMW dealer organisation. A key reason behind the significant increase in volumes over the previous model was the global market launch of the new 3 Series within the space of just two months. Whereas right-hand-drive variants of the fourth-generation car only went into production some six months after the left-hand-drive model, the E90 was available with RHD from the outset.

In order to achieve a start-up curve as steep as that of the E90, a product has to have reached an advanced stage of development well before the start of series production. For this to be the case, all the process partners in development and production - plus the external suppliers - need to be integrated in a timely manner into the product development process. Plus, safeguard planning and meticulous checking strategies were carried out at the plant early on in order to ensure fault avoidance in the electrics and electronics, and that also gave the start-up a solid platform. Additional intensive vehicle tests as early as the pre-series phase also helped to monitor the interplay between complex electronic components. The approach of first checking smaller units

Body Shop: the section of the line leading up to the assembly of the bonnet and tailgate.

Building on the front end.

2004 The BMW 3 Series Saloon E90 goes into production

before moving on to larger assemblies proved to be particularly efficient. In this way, it is possible to identify precisely at which point in the interplay a fault has occurred. And this equates to valuable time savings when it comes to establishing as quickly as possible what changes need to be made.
This lead plant approach – where one plant ensures that the product is ready for production before the process also begins at the other sites – ensures that the necessary coordination between plants and development areas only has to take place once. The other plants can then focus their resources more effectively on bringing the model variants that only they are building (in the case of Regensburg, the 3 Series Coupé and 3 Series Convertible) up to series production readiness. Just six months after the new Saloon went into series production, the 3 Series Touring also started rolling off the assembly line in Munich. This achievement broke new ground for the plant in terms of speed. And only since the required buildings and technical underpinnings were put in place in 2003/04 has it been possible to produce the Touring and four-wheel-drive variants in Munich.

Michael Eichinger (Finishing Area)

BMW is a great employer, and working here is enjoyable. I get the feeling from my team that they like coming to work. Many of the guys have to travel a long way to get here and some of them are on the road for up to one and a half hours. I have more than 200 people in my team from several different nations, and that means many different characters. We have very good people here, and they always have – and embrace – the opportunity to learn new skills. Standing still would not be a good thing. It is important to work out changes together with the people affected. This not only keeps them closely involved, it also means they can use their expertise to good effect. And that often helps to avoid the occurrence of faults from the outset. Treating people honestly and fairly is what it's all about. My experience at BMW is that everybody is employed to make best use of their own personal strengths. Commitment and a drive to move things forward are recognised and rewarded. This plays a major role in our success. Not everybody was that pleased when associates were increasingly being switched around between plants a few years ago. However, we in Munich have also enjoyed the benefits. Each new person brings new skills with them and we're all constantly learning – so there has clearly been a lot gained. Returnees from locations outside Germany also settle back in quickly. The special thing about the Munich plant is its close proximity to the Research and Innovation Centre. For me it is a big advantage to be able to get together with colleagues in the development department in just a short space of time. This allows us to make a lot of progress together.

The fourth-generation BMW 3 Series Saloon (model year 2002).

Sheer Driving Pleasure "made in Munich": cars

In May 1998 the fourth-generation BMW 3 Series (E46) went into production, initially in four-door Saloon guise. The 3 Series had remained true to its roots in terms of dimensions, measuring just four centimetres more in length and width than its predecessor and only 12 centimetres more than the first 3 Series from 1975. The hallmark double kidney was now integrated into the bonnet and combined with the twin circular headlights behind clear covers to define the expressive character of the new 3 Series face. The chassis was a new development, with some axle components made from aluminium and others – like the front and rear axle carriers – from high-strength steel. This meant they were lighter than their predecessors in the

The second-generation BMW 3 Series Compact was launched in the spring of 2001.
The picture shows the model update of March 2003.

outgoing model. A stand-out safety feature of the new car was the ITS head airbag, a tubular structure concealed in the door frame next to the driver and front passenger which provided significantly better head protection than previous systems.

The new 3 Series Compact took to the stage in 2001, its new interpretation of the characteristic BMW twin headlights lending it an unmistakable front end. The 316ti was the world's first series-produced car to be fitted with a Valvetronic engine. Developed by BMW, this technology was the first to allow variable lift control of the intake valves. Engines with Valvetronic do not require a throttle valve to control engine load, something which becomes apparent under partial loads – i.e. low and medium engine revs – through noticeably lower fuel consumption.

BMW unveiled a spectacular new roadster at the 1999 Frankfurt Motor Show. At first glance an apparent nod to the legendary BMW 507, the

Cockpit of the BMW Z8.

uncompromising form of the BMW Z8 is every inch the pure, classic roadster. Breathtakingly beautiful and classical in its proportions, yet based on innovative technology such as an aluminium space frame body, it breathed new life into a historical legend.

The assembly of the Z8 began in January 2000 – at the Munich plant. And for good reason. More than ten years earlier the plant had provided evidence of its expertise in the field of small-series production by building the Z1, now considered a classic. A separate area was created for the Z8, set apart from 3 Series production, in the former Building 159 pilot plant. A total of 31 cycles went into turning the painted bodyshell (supplied by the Dingolfing plant) into the finished product. The ergonomics of the individual workplaces in the new area were taken into account as early as the planning stage. Variable swivel mechanisms and height-adjustable transportation systems allowed strenuous overhead work to be avoided. Ability to work in a team – as well as qualifications – played a critical role in the

The BMW Z8: 5-litre V8, 400 hp/294 kW, 0-100 km/h in 4.7 seconds.

The BMW Z8 prior to wheel assembly.

selection of the around 100 production associates in the Z8 assembly section. Preparation and training for these associates took place in a training programme lasting several weeks and specially designed for the assembly of the Z8. For the associates, each cycle meant having to master working processes lasting approximately 50 minutes. This flexible team and production concept had already proved itself by the start of series production. Demand initially rose to twice the planned capacity, and yet no customers had to wait an excessive amount of time for their Z8. By mid-2003, production figures for this dream roadster (including the Alpina Roadster V8) had reached 5,700 units.

Just when people were wondering aloud if there was any way you could still improve on the ultimate sports saloon (the E46 3 Series), just such a car arrived in the shape of its successor, the World Car of the Year 2006. From March 2005 to early 2012, the fifth-generation 3 Series - the E90 - wrote the next chapter in this outstanding success.

The BMW 3 Series E90, World Car of the Year 2006.

The BMW 320si – the roadgoing version of the BMW 320si WTCC – was built at the Munich plant. Andy Priaulx drove the race-trim car to the 2006 FIA World Touring Car Championship (WTCC) crown.

BMW 3 Series Touring (E91).

Numerous technical highlights such as Active Steering and Active Cruise Control (ACC) made their 3 Series debut in the E90.
September 2005 saw the debut of the BMW 3 Series Touring E91, the first lifestyle estate car to be built at the Munich plant.
For the first time, a BMW 3 Series Touring was developed from start to finish alongside the Saloon. Essentially identical to the Saloon from nose to A-pillar, the rear seat and boot areas of the Touring slipped organically into the template of a sporty and almost coupé-like silhouette. Here, the side window surfaces played the pivotal aesthetic role. The roofline fell away smoothly to the hatchback rear and the shoulderline rose slightly, creating the small window surfaces reminiscent of a coupé. The broad C-pillar guided the eye back to the Touring's rear axle and sporty rear-wheel drive.

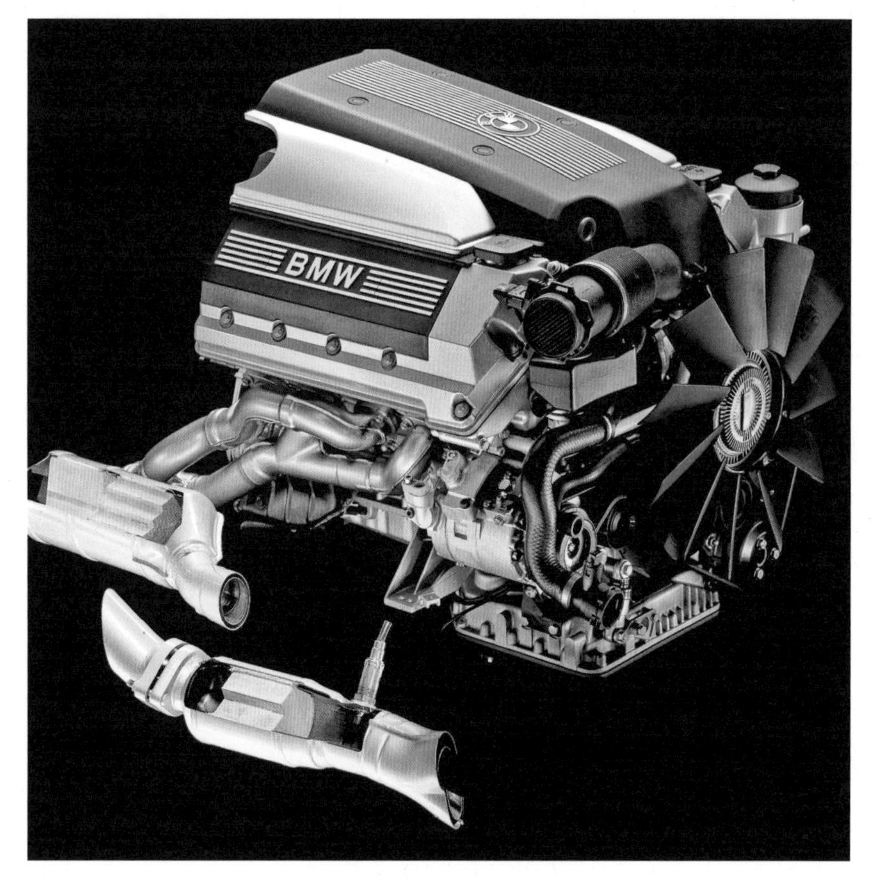

The V8 S62 was fitted in the BMW Z8 and in the BMW M5 (E39).

Sheer Driving Pleasure "made in Munich": engines

The Munich engine factory has kept an unparalleled product firework burning since the late 1990s. The fuse was lit in 1999 by what was, at the time, Germany's largest eight-cylinder diesel engine with common-rail direct injection (the M67) – the first ever diesel powerplant to be built in Munich. Available for the 7 Series, it caused quite a stir as the first diesel V8 in the luxury segment, developing 238 hp/175kW of output and peak torque of 560 Nm. The electric turbocharger adjustment was an innovation fitted exclusively in the 740d.

The burning question in the run-up to the presentation of the M3 Coupé at the Geneva Motor Show in the spring of 2000 was whether the 3.2-litre

The S54 in-line six-cylinder engine.

six-cylinder engine taken from its predecessor – and developing 312 hp/ 236 kW at 7,400 rpm – still had anything left in the tank. It certainly did. A combination of the high-revving concept pushed to its next level and numerous innovations in the cylinder head allowed the S54 to develop 343 hp/252 kW at 7,900 rpm, with maximum torque reaching 365 Nm at 4,900 rpm. The E46 M3 Coupé accelerated from 0 to 100 km/h in just 5.2 seconds.

The eight-cylinder M62 petrol engine (3.5 – 4.4 litres), production of which began in 1996, was extensively reworked in 2001 and also fitted with the innovative Valvetronic valve control system (N62). The 5-litre eight-cylinder unit (S62, from 1998) fitted in the M5 (E39) and Z8 was succeeded in 2004 by a ten-cylinder engine (S85). Up to then normally reserved for racing cars and automotive exotica, a high-revving V10 celebrated its crossover into the world of series-produced BMW Saloons in the M5: ten cylinders, 5-litre displacement, 507 hp/373 kW worth of output, 520 Nm of torque and a maximum engine speed of 8,250 rpm. In simple terms, this was a power merchant par excellence.

An engine mechanic adjusts the camshafts.

The M73 12-cylinder engine in the BMW 760Li, 2001.

Series production of the 12-cylinder engine (N73) began in January 2003. This powerplant was fitted in the BMW 7 Series E65 (5,972 cc, 600 Nm, 445 hp/327 kW).

It offers just the unbeatable smoothness, elasticity and comfort that customers rightly expect of a V12. From July 2005 to September 2009 the Munich engine plant produced a new generation of in-line six-cylinder petrol engines (N52) on a totally new assembly line in Hall 140. In the interests of reducing weight, the engine saw BMW using magnesium for the first time in volume production. 30 percent lighter than aluminium, this material was selected for the crankcase, crankshaft bearing and cylinder head cover. Valvetronic was another new feature of the six-cylinder engine. In the 2005 edition of the annual Engine of the Year Awards – the world's most important engine competition (otherwise known as the "engine Oscars") – the 3.2-litre powerplant in the M3 took the honours in the 3.0- to 4.0-litre class for the fifth year in succession. However, the main award ("Engine of the Year") went to the 5.0-litre V10 engine fitted in the M5 and M6.

BMW GROUP
Werk 1.1 Tor 1

DRIVING LUXURY.

Modernisation with production running and series-built legends

Leadplant for the BMW 3 Series and a beacon for electric cars. The Munich plant is expanding its portfolio and restructuring its production. A new state-of-the-art paint shop is being built and the assembly division is undergoing the biggest rebuild in its history. But the face of the BMW home plant remains truly unmistakable.

2010 – 2020

The assembly structure VFlex, 2006. Complex parts are pre-commissioned in the form of "engine sets", transported to the main assembly line and linked up with the workpiece carrier. Highly qualified associates piece the engines together. The aim of this line concept with standardised robot assembly cells is to achieve maximum flexibility, thus allowing assembly of as many engine families and variants as possible on a single line.

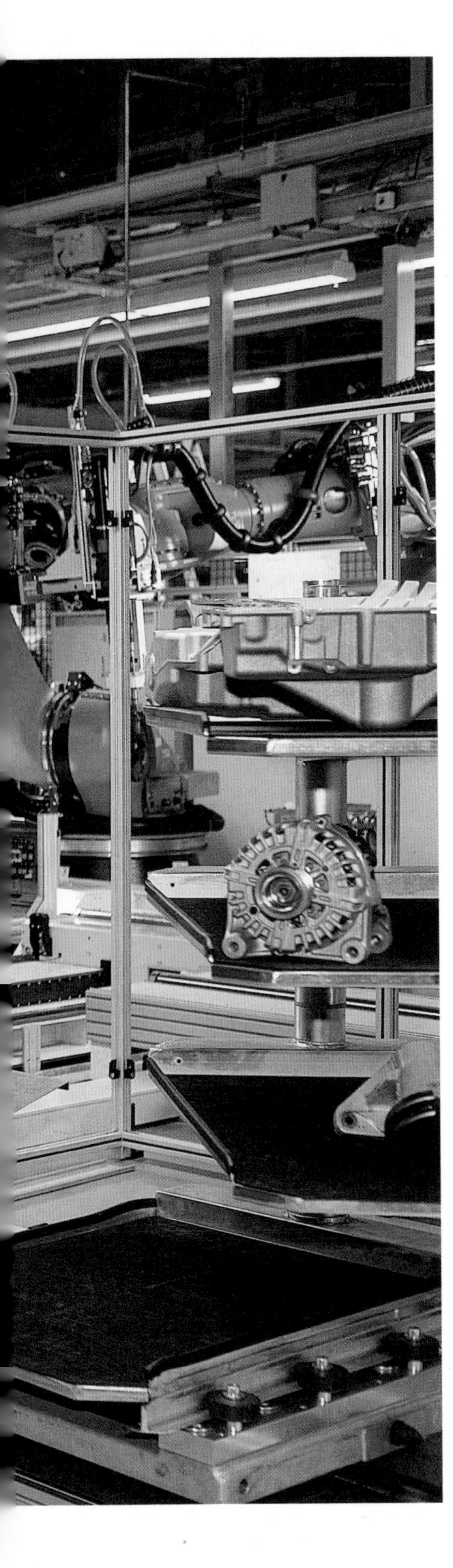

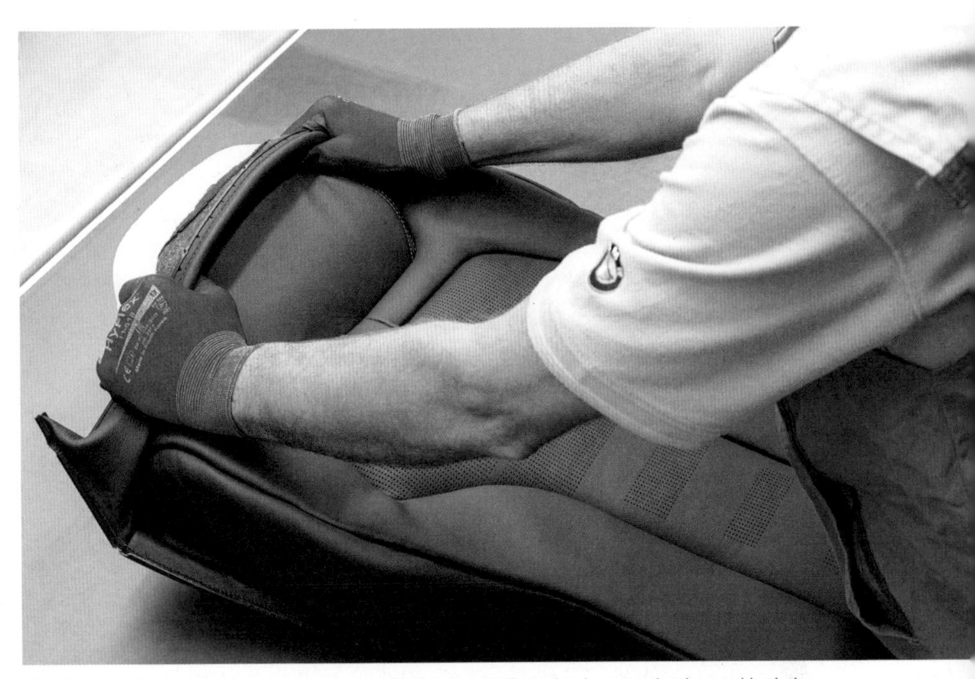

The Seat Production Centre pursues a consistent policy dedicated to lean production and logistics processes. The Agamus corporate consulting firm honoured this commitment with a Lean Award in 2015.

A complete plant

Despite the structural changes of recent decades, the Munich plant – which today has a workforce of approx. 7,700 – continues to boast a unique spread of automotive production expertise. The entire vehicle manufacturing process chain – the press shop, body shop, paint shop and vehicle assembly – links together at this one site. And that's not to mention the engine production halls, seat manufacturing facility and laboratory that are also located here. The laboratory is part of the quality assurance system for new model launches and serial quality monitoring. It plays an especially important role in the process of analyzing storage cells for high-voltage batteries in electric powertains.
Of course the Munich plant does not do everything itself. Indeed, it is integrated into a global development and production network which also includes suppliers. The company's relationships with many of these stretch back years. A particularly important partner is the BMW plant in Regensburg, which supplies Munich with all the doors and lids for the body shell, for example.

More than just hoses and water

A car plant can, in some respects, be compared with a city. For instance, both offer restaurants, shops and a fire service. The fire service at the Munich plant performs the main tasks preventative fire protection and fire fighting, including technical assistance. The plant fire service is kitted out with state-of-the-art vehicles and all the equipment it needs to do its work. Preventative fire protection involves supervising licensing procedures for work carrying a risk of fire, conducting checks on the functioning of fire extinguishing equipment, fire alarms and fire detection systems, and providing advice on new constructions and conversions or changes in a building's use. If required, the plant fire service will – in the interests of good neighbourhood relations – support public fire services with special-purpose equipment and/or vehicles, as far as its capability allows.

The Munich factory is the lead plant for the BMW 3 Series. Following the highly successful launch of the sixth generation (F30) at four plants beginning in 2011, Munich is still responsible for coordinating quality assurance measures in production. Another aspect of the lead plant function is the integration of model-year measures (such as new equipment features) in close cooperation with the other BMW 3 Series plants in Regensburg, Rosslyn/South Africa and Tiexi/China. The F30 is the last 3 Series for the Regensburg and Rosslyn plants. But a new site is joining the ranks of the BMW Group's global production network.

The Munich plant and its team, therefore, have a new and important task. As the lead plant, they are partially responsible for setting up the new production site in San Luis Potosí, Mexico, in order to prepare for the seventh generation of the BMW 3 Series. Before the first BMW 3 Series from Mexico

Health Management

The BMW Group has developed a multifaceted programme devoted to enhancing the health and productivity of its associates by encouraging them to remain physically active. Hands-on activities and campaigns encourage employees in all areas to participate and get involved.

Services offered range from traditional preventive measures, such as health check-ups and a fitness studio located near the plant, through temporary activities like 'fit on the work bus' to bicycles provided for work-related errands at the Munich site.

was produced and the plant had its ceremonial opening in June 2019, hundreds of new colleagues from the Central American country travelled to Munich for general and assembly training. In return, the setting-up of the Mexican was shaped by almost as many specialists from all disciplines at HQ and at other plants.

In 2021, the BMW Group's production network (all vehicle, engine, component and assembly plants combined) comprises 31 locations in 15 countries and has a global sales network in more than 140 countries.

Success by subscription: the Munich plant's BMW 3 Series line-up

In its February 2012 edition, *auto motor und sport* welcomed the new BMW 3 Series Saloon (F30) enthusiastically, underlining that "with its dynamics and agility [...] it points right into the sports driver's heart. People who develop such cars really must be appreciating Sheer Driving Pleasure". The sixth generation of the BMW 3 Series Saloon, the world's best-selling premium car, set new benchmarks in terms of sporting prowess, elegance and comfort. The Touring vehicle, which is almost as popular in Europe, appeared in the summer of the same year.
In November 2015, the first plug-in-hybrid of the BMW 3 Series Saloon went into production in Munich. In pure electric mode, the BMW 330e covers a range of up to 40 km without producing any local emissions. This new model variant was another success: soon after its market launch, one in six vehicles from the Munich plant was an electric BMW 3 Series.

40 years as the original

With a total of over 14 million vehicles delivered to customers since its initial introduction, the BMW 3 Series is the biggest selling BMW of all. Accounting for around 25 per cent of total BMW vehicle sales, the BMW 3 Series Saloon and Touring together represent the BMW brand's most successful model range. The Munich plant celebrated another production anniversary in September 2015, when the BMW 3 Series no. 10,000,000 rolled off the Munich assembly line (a figure that includes all six generations built over 40 years on all BMW production sites).

40 years of BMW Series 3, May 2015, BMW Museum: a very personal anniversary for many associates.

A BMW 320d EfficientDynamics Edition Touring, photographed in July 2015 (updated edition). The most fuel-efficient Touring model takes pride in consuming just 4.0 litres/100 km (standard tyres, eight-speed automatic unit).

There was no end in sight for 3 Series fans. In October 2018 the seventh generation of the 3 Series Saloon was presented at the Paris Motor Show. The Munich plant's employees were able to watch the vehicle's presentation at the trade fair via a live broadcast in a large hall at the northern edge of the plant. There, BR-TV sports presenter Markus Othmer hosted the enjoyable world premiere, including Mexican waves, and asked the then Production Director Oliver Zipse and Plant Manager Robert Engelhorn to unveil the vehicle. Afterwards, the stage was set for selfies and team photos with the home plant's new star.

The 3 Series Touring and the 3 Series Plug-in Hybrid were also set to be revised, of course – both in summer 2019. At the same time, a plug-in hybrid variant for the 3 Series Touring was also announced for the first time. It was launched in summer of the following year.

The BMW 3 Series line-up was completed with the BMW M3 at the Munich plant from autumn 2020. 35 years after the first M3 made its debut, the plant's team were delighted to land the contract to produce the latest generation of BMW M GmbH's high-performance car. At the start of production,

An icon with many fans in the Munich plant: the BMW M3 (2020). (top)

Since November 2015, the BMW 3 Series saloon has also been offered as a plug-in hybrid. (bottom)

Plant Manager Robert Engelhorn said: "The car is absolutely iconic and has many many fans at the plant – me included. "The team benefited from their wealth of experience with the previous generations of BMW M vehicles when integrating the new M3 into the home plant's production processes. For example, the independent assembly process for the CFRP roof of the M4 Coupé, which was now produced at the Dingolfing plant instead of in Munich, could also be adopted for the new M3. On the other hand, seat production required entirely new hand movements for making the newly developed M Carbon bucket seats. The hood and side panels of the M3 are also handmade, as well as being assembled manually.

The Coupé, another 3 Series variant, is likewise produced at the Munich plant. At the start of the fourth generation, it was rebranded into the newly founded 4 Series. With the new edition of the sporty two-door car in 2020, production of the 4 Series Coupé was put in the capable hands of the Dingolfing plant, where the 4 Series Gran Coupé was already rolling off the production line.

BMW M4 Coupé. Four generations of mid-size M Coupés have blended motor sport genes and uncompromising everyday usability within an emotionally rich overall concept.

The bonding plant for the carbon roof of the M4, which is part of the paint shop. This is the first time the Munich plant has handled production of a lightweight design concept, bringing together an aluminium bonnet and side panels, CFRP roof, and lightweight tailgate in this combination.

Home of the most powerful letter in the world…

.... Meaning, of course, the M of BMW Motorsport GmbH, which has consistently had its cherished performance cars produced at the home plant over the affiliate company's 50-year history.

Having produced the BMW 4 Series Coupé for many years, this provided an excellent basis for also producing the most dynamic car in the segment: the BMW M4 Coupé. After its predecessor, the fourth-generation BMW M3 Coupé, had more than 40,000 units produced, the BMW M4 Coupé took up the mantle from 2014 onwards. Being given the name "M4", it also clearly references the model series that served as the starting point for the new M model.

In August 2015, a year on from the launch of the BMW M4 Coupé, BMW M Division presented an initial preview of a high-performance model for use on the road and, above all, on the race track: the BMW Concept M4 GTS. While the BMW M4 Coupé embodies the ideal combination of motor sport

genes and unrestricted everyday usability, the BMW Concept M4 GTS previewed an emotionally powerful and exclusive special model, which was conceived with an eye for trailblazing technology and a keen focus on the racetrack.
Less than a year later, the BMW M4 GTS was released in an edition of just 700 units, which were also produced with great passion at the Munich plant.

"DTM-Event" at the Munich Plant with a 2014 season DTM M4: "burnout" on the main traffic axis of the plant. In the 2014 season BMW secured the constructors' title as well as the team title, BMW driver Marco Wittmann won the drivers' title.

On 27 February 2014, DTM Driver Martin Tomczyk (pictured with Plant Manager Hermann Bohrer) drove the very first series production BMW M4 Coupé off the assembly line. The decision to build the M4 in Munich was among other factors motivated by the synergies that arise when integrating the high-performance sports car into the production processes of its donor model.

Meeting of Legends

The BMW M4 Coupé and the BMW M3 Sport Evolution of the E30 generation. Following a 23-year break, the home plant has returned to producing a BMW M model again since February 2014. Traditionally, the myth of BMW M is closely connected to the Munich site. Between 1986 and 1991, BMW's plant in Munich built 17,970 units of the first-generation BMW M3, of which 786 were convertibles. Many people who are still working at the plant today were involved in the production of the BMW M3 E30.

New investments for the lead plant of the BMW 3 Series...

Prior to the start of production, each generation of cars requires new investment in the plant. Before starting the sixth generation of the BMW 3 Series Saloon in October 2011 until mid-2017, the BMW Group invested half a billion euros into its home plant – once again a clear commitment to its 'oldest' site.

To ensure the utmost precision and efficiency, new production structures were created for all the manufacturing areas. The new products benefit, for example, from a new large press. At 17 strokes per minute, this installation counts among the most advanced in the world. It achieves throughput of 600 tonnes of steel per day and develops press forces of 650 to 2,500 tonnes. The new production infrastructure also ensures the highest possible economy of space and optimal material flows.

To produce the new models, a completely new production building for the body shop was constructed. Here, state-of-the-art robots were installed,

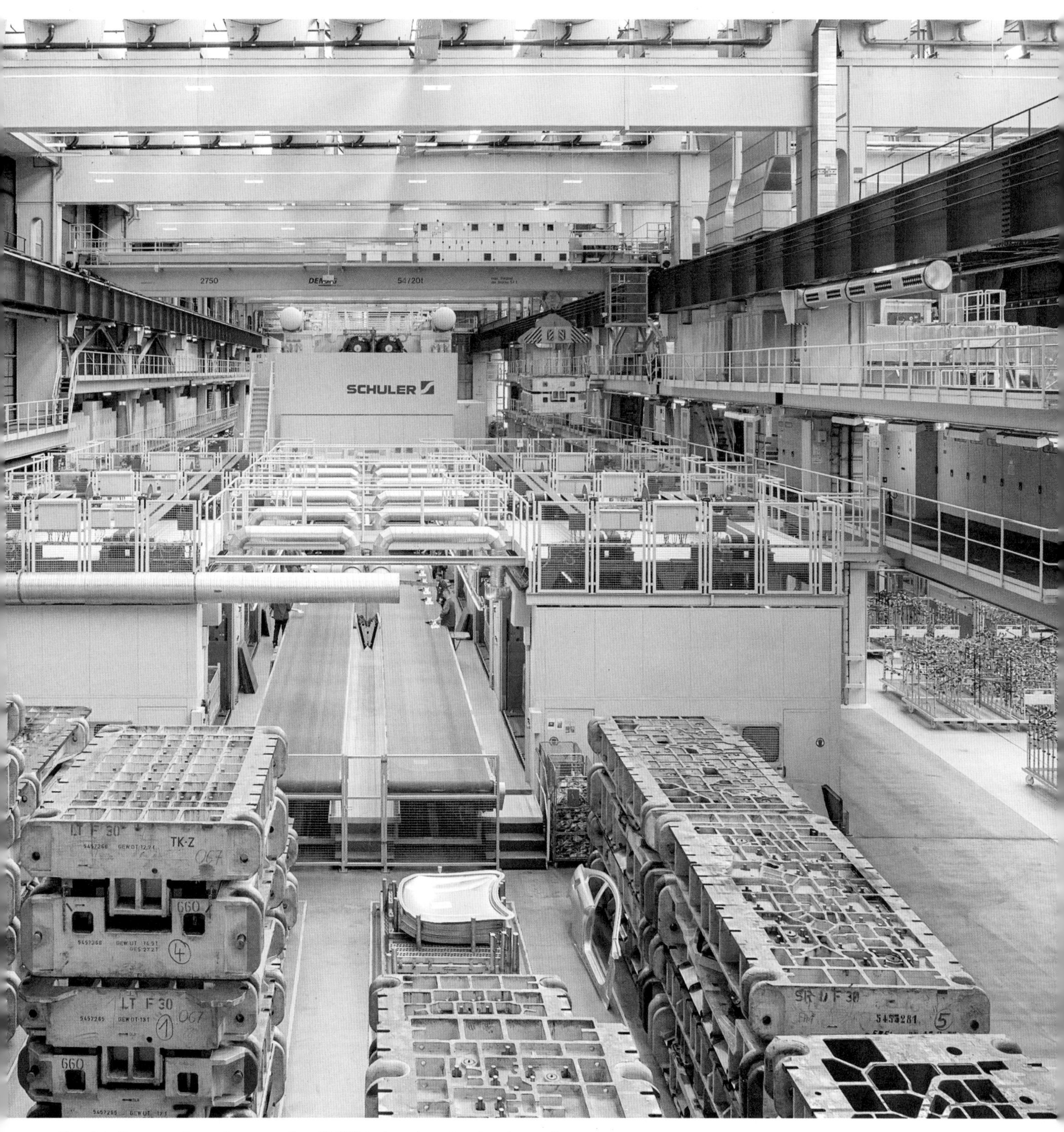

View into the press shop, where more than 32,000 body parts are produced every day.

Precise robots, experienced hands and eyes: Together, they ensure premium quality in body construction.

some of which had previously been operating at the Leipzig plant. Synergies such as this make a noticeable contribution to cutting costs and ensuring the sustainable use of resources. In addition, latest-generation laser robots and advanced bonding robots are being used. This modern, low-temperature process is now being used for around twice as many operations compared to the predecessor models, in part replacing conventional spot welding.

The innovative technique significantly improves sealing and corrosion prevention. The immediate customer benefit is a stiffer body, significantly enhanced driving dynamics, comfort and safety.

When preparing the plant for the current products, great care was also taken, for example, to protect local residents from noise generated by production. Noise abatement measures include the use of innovative silencers,

ventilation equipment and sound absorbing cladding, as well as optimised transport logistics. Paint shop odours are avoided thanks to state-of-the-art filtering and regenerative thermal oxidisers.
Beyond this, all production processes are subjected to continuous monitoring, which includes assessing the impact on local residents. Since July 2015, intra-city inbound logistics for the Munich plant includes a 100% electrically operated truck – a 40-tonne truck from the Dutch manufacturer Terberg. For a start, this truck commutes eight times per day between the SCHERM GRUPPE logistics centre and the Munich plant.

Plant manager Hermann Bohrer, Bavaria's Minister of Economic Affairs Ilse Aigner and representatives of the logistics partners present the first electric truck for material transport in public road transport in July 2015. (top)

The green roof of the new body shop not only has a sound-insulating effect, but also around 160,000 bees have their home here. (bottom)

Assembly of the modular engines, 2014.

… and the lead plant for petrol engines

Towards the end of the first decade in the new century, the engine production plant in Munich was facing another big change: In the early 'noughties', due to the high demand for six-cylinder petrol engines, the existence of two production sites (Steyr/Austria and Munich) still made a lot of sense; after all, this powertrain was fitted in all model series of the BMW brand and accounted for over half of all of the brand's engines. However, the trend towards smaller cubic capacities and a lower number of cylinders at an equal or even increased engine output (and lower fuel consumption) did not remain limited to diesel drives. An engine output which petrol engines without turbo-charging could only achieve with six cylinders, was now easily feasible with four-cylinder engines – thanks to turbo-charging. At the same time, it became obvious that it would soon be possible to match the performance of some naturally aspirated four-cylinders with turbo-

Three and a half years after the start of production, the one-millionth four-cylinder Twin Power turbo petrol engine left the engine production plant. For a while, it was fitted in nine different BMW model series of five different output ranges up to 245 hp/180 kW.

charged three-cylinder drivetrains. BMW played a major role in shaping this trend with its Efficient Dynamics strategy. Increasingly challenging fuel consumption and/or carbon emission targets issued by numerous international regulation bodies made it irreversible. This is why the BMW Group decided to limit the future production of six-cylinder petrol engines to the Steyr site. However, the discontinuation of the production of this particular engine range was not meant to reduce the size of operations in Munich; on the contrary, the long-standing reference value in production of approx. 300,000 units annually was supposed to more than double in the long term, thanks to new three- and four-cylinder petrol engines whose series production was launched in two steps in 2011 (four-cylinders N20, N26) and 2013 ('modular engines' three-cylinder B38, four-cylinder B48).
As in the car factory, space is a scarce commodity at Munich's engine plant. Therefore, the new drivetrains were to be manufactured primarily in the areas previously occupied by the six-cylinder engine production. Consequently, the production of the three- and four-cylinder engines could only commence after the dismantling of the six-cylinder installations. In a joint show of strength, the engine plant's management, HR and the works council managed to ensure continuous employment for all associates affected by the conversion phase – despite the fact that production output had meanwhile decreased to some 70,000 drivetrains per year.
The reward for the great efforts was that the Munich site was extended to become the lead plant in the BMW Group's international engine plant network for the production of petrol engines. To make this possible, the company invested approx. 50 million euros in the Munich-based plant in 2013 alone, about two thirds of it earmarked for the new production line for modular engines.
The standardised engine module allows the BMW Group to manufacture three-, four- and six-cylinder petrol and diesel engines of different output ranges on a single assembly line. The new engine range is based on a uniform in-line construction concept. Thanks to the high number of common parts, significant cost savings can be achieved in development and production.
Engine expertise from other sites was applied when setting up the new modular engine production facility as well: Both the setup and the launch of the production lines took place at the same time – between autumn 2012 and late 2013 – at the sites in Munich, Hams Hall and Steyr. The three engine production plants each optimised their own launches, coordinated best practice solutions within the entire network and installed mostly identical production facilities, despite differences in combustion technology.

Leak test in the Munich engine plant, 2014.

Another common construction principle of the modular engines is the single displacement size of 500 cc per cylinder. The particularly low-friction basic engine made completely of aluminum is thermodynamically optimised, very lightweight and quiet in operation. One (three-cylinder) or two (four-cylinder) counter-rotating balance shafts offer supreme engine smoothness throughout the entire speed range. Other common features include standardised interfaces, such as the identical engine mounts and the connections to the cooling circuit, the intake and exhaust gas system as well as the heating and the air conditioning system. All petrol engines are equipped with an identical turbocharger integrated into the exhaust manifold. In this setup, not only the exhaust manifold, but also the aluminium turbine casing is liquid-cooled. The short exhaust flow paths to the charging system ensure agile response characteristics while the very short heat-up phase reduces internal system friction and thus fuel consumption. As the catalytic converter and the electrically controlled boost pressure control valve are positioned close to the drivetrain, the petrol engine's emission performance is very favourable.
The Munich engine plant is still the one and only site to manufacture V engines within the BMW Group. Munich-made engines power vehicles of the brands BMW, MINI and Rolls-Royce.

In addition to the N20, the modular engines and V8 and V12 petrol engines, the Munich engine plant still manufactures a diesel engine, indeed a very special one: in March 2012, series production for the "Top Diesel" 50d started. Pictured is the engine of a BMW M550d xDrive. Three turbochargers with variable charging and variable turbine geometry plunder 381hp/242kW from 2,993cc displacement.

Born electric

Quite naturally, the era of electric mobility has dawned in the Munich plant as well. Pilot assembly of the BMW i3 and the BMW i8 started early this decade in the former pilot plant, Hall 159, which had already hosted (series) assembly of the roadsters BMW Z1 and BMW Z8. Here, the series manufacturing processes were tested for the BMW Group Leipzig plant, where production of the i3 started in 2013 (and series production of the i8 in the following year). The MINI E, the first BMW Group small series production car to be powered purely by electricity, also received its drivetrain and high-voltage batteries in Hall 159.

Resource-efficient and innovative: the new paint shop at the Munich plant.

The future of painting

At over 200 million euros, the largest individual project within the modernisation programme is the construction of a completely new paint shop. The facility comprises a space of about 13,500 square metres and, since its completion mid-2017, is one of the most innovative paint lines in the industry, in terms of both profitability and the efficient use of resources.
Innovative, six-axled robots offer considerably more leeway when it comes to coating technologies, allowing for greater flexibility in responding to customers' special wishes, such as individual colours or matt paint coats.
The facility will apply the innovative and highly efficient IPP (Integrated Paint Process) technology. Compared to conventional procedures, this paint process completely omits the process step of applying and baking of a primer, whose functionality is transferred to one of the two base paint coat layers. In the wet-on-wet procedure, the first of the two layers assumes all functionalities and characteristics of the primer while the second guarantees visual characteristics, such as colour, lustre and depth effect.

Just as before, the base coat is covered with a clear coat, making sure that the IPP technology meets the same high standards with regard to the paint coat's look and protective function as conventional paint processes.
The introduction of the IPP will be another important step towards sustainable production of the plant situated in the heart of Munich: compared to the current paint shop, those natural gas consumption and exhaust emissions at the new facility are expected to decrease by about half (48 percent each), and energy consumption by over a quarter (27 percent). The overall energy saving achieved corresponds to the average annual energy consumption of 4,000 two-person households. Emissions of volatile organic compounds (VOC) will be reduced by about a third (35 percent).

In May 2015, the new paint shop at the Munich plant was opened by plant manager Milan Nedeljkovic, Ilse Aigner (Bavarian Minister of State for Economic Affairs), Production Board member Oliver Zipse and the Chairman of the Works Council Manfred Schoch. (top)

Precise and efficient: Ink application with IPP technology. (bottom)

In 2018, the current BMW 3 Series saloon was presented to the trade press at a race track in Portugal.

New 3 Series, new technology

Since 2015, the Munich plant has been preparing to start production of the seventh generation of the BMW 3 Series. To accommodate the numerous innovations and increased complexity of the new model, the body shop and assembly were expanded in addition to the new paint shop. The digitalisation and density of networked systems are increasing in all areas. The highly complex safety, driver assistance and connectivity systems in the new BMW 3 Series require equally complex assembly, testing and inspection facilities.

Straight after starting production of the new 3 Series generation in autumn 2018, the plant began preparing to produce the new BMW 3 Series Plug-In Hybrid in mid-2019. The BMW 330e also has certain conceptual changes compared to its predecessor, which have noticeable consequences for production. The Munich plant is perfectly equipped to meet the increasing demand for electric vehicles. If needed, almost every third BMW from the home plant could be a 3 Series plug-in hybrid. With its predecessor, the maximum capacity was only around ten percent of daily production.

The range of models at the Munich home plant in 2019 includes five vehicles with consistently high demand: the new BMW 3 Series Saloon with conventional drive and as a plug-in hybrid, the new BMW 3 Series Touring

Front-end assembly of the BMW 3 Series (G20).

A fully automated future: expanding the body shop

As the developers are striving for even greater dynamics and reduced vehicle weight, while at the same time further improving safety for the occupants, this also increases the demands on modern body construction. In order to be equipped for the next generation of the BMW 3 Series, between 2016 and 2018, the Munich body shop was expanded by around 24,000 square metres.

As the saying goes, out with the old, in with the new. The 'old' here was a building that had been used as a press shop for decades, complete with basement and supply tunnel. During the demolition, which was carried out with centimetre-perfect precision, concrete and soil contaminated with mineral hydrocarbons had to be professionally disposed of. A temporary façade with an area of around 3,000 square metres was erected at the directly adjacent storage facility to protect the production there from dirt.

The new building was connected to the existing production areas of the body shop while it was still running. Step by step, old and new co-existed. During this phase, it was possible to work on both the new facilities and the building effectively in parallel, thanks to a high degree of coordination. This way, important sub-areas such as conveyor levels and a tunnel connection that were needed at a very early stage could be finished in advance.

Another special feature of the project was the position of the body shop: directly on the edge of the boundary of the plant with Dostlerstraße. For that reason, the road had to be partially closed and the bus stops for the works bus service had to be moved. Therefore, the aim was to minimise the construction area needed for the site equipment and logistics, as the building project already took up the entire area between the existing buildings and the plant boundary.

also launched in 2019, as well as the BMW 4 Series Coupé and the BMW M4. At the beginning of the year, plant manager Robert Engelhorn announced that "Production volume in 2019, at around 230,000 vehicles, will be one of the highest in the plant's almost 100-year history."

New investment in e-mobility

A good year after the new paint shop went into operation, the next modernisation steps had already been announced. Having got the go-ahead to produce the first all-electric Gran Coupé – the BMW i4 – the Munich plant invested around 200 million euros in buildings, production facilities and logistics systems from 2019, to realise the i4 series production by 2021.
When this was announced in December 2018 it had already become clear that this was to be the largest conversion in the plant's history – and would happen while production continued. Integrating the BMW i4 requires extensive action, especially in constructing the body and assembling the vehicle. The vehicle's body differs significantly from the architecture of the models previously produced at the Munich plant. For instance, the BMW i4's high-voltage battery requires an almost completely independent floor assembly including the rear. Therefore, the planning specialists now have to design complex interlinked production lines so that the 1,000-plus robots are not only able to seamlessly produce the bodies of the BMW 3 Series, 4 Series and the M4, but also the BMW i4's special body.
Integrating the assembly of the BMW i4 is also a challenge, not least due to the battery. In order to build the high-voltage battery into the vehicle, the planners have to accommodate extensive and space-consuming conveyer and systems technology, on the already cramped Munich factory floor.
The plant's logistics and material supply also comes under pressure from the BMW i4, as numerous i4 components differ from conventional vehicle components. This in turn means additional component variants, which exponentially increases the goods flow to be managed.

New equipment for the side frame of the BMW i4 was integrated into body shop in summer 2020.

The plant management circle with the BMW i Vision Dynamics - the concept car for the later BMW i4.

At Riesenfeldstraße 87, a new logistics centre for the high-voltage storage systems of the BMW i4 was installed.

A plant tour on its way along the production mile in the BMW Group Munich plant.

Showcase

BMW Welt boosts interest among customers, visitors and brand aficionados in plant tours – an extremely "real" way of experiencing BMW in action. The opportunity to enjoy an exclusive glance behind the scenes and to discover exactly how Sheer Driving Pleasure is conceived is available not only to customers collecting a new car. A tour illustrates in detail how individual vehicles are tailored to customer requirements at the home plant of the BMW Group. Visitors experience how parts are manufactured from tonnes of steel roll, joined to bodies and painted. They can see how an engine is created, witness the assembly process of marrying the powertrain and body, and observe the numerous quality tests being carried out. Along the "production mile", plant tour experts give the visitors unique insights into the manufacture of premium cars and engines. All of which makes the BMW Group Munich plant a showcase location for BMW production.

Finished cars, seen from the production mile.

Contrasts

The buildings within the Munich plant offer captivating contrasts between a historical architectural core and modern production buildings constructed in the image of the BMW Corporate Identity. In the northern section of the plant in particular, there are several buildings still surviving from its early days. On Dostlerstraße, the plant is shaped as much by the modern Gate building 34 as the older Building 71, which housed the plant management between 1982 and 2018.

It may be the company's original plant on the one hand and a cutting-edge production facility on the other, but that is no contradiction for the workforce at Munich. Generations of associates and management have taken the right approach in dealing with the apparent disadvantage of a comparatively compact surface area. They have seen optimum process management and

The brick building on Dostlerstraße was the headquarters of the plant management until 2018. Next to it, the new body shop.

innovative solutions as the most effective answer to the structural restrictions imposed on an industrial company by the growth of the city around a plant. Displaying a readiness to repeatedly tackle supposedly insurmountable challenges, the plant has managed to emerge ultimately unscathed from the fissures and breaks in its long history. Indeed, change has been a constant companion in the plant's evolution from a manufacturer of aero engines to the key plant in 3 Series production and becoming a beacon for e-mobility.

The new building in the centre of the plant combines assembly, planning and plant management functions.

At Gate 1, buildings from different decades nestle together.

The new high-bay warehouse stands on stilts. The main logistics route still runs underneath it.

View from BMW Welt to the plant's wall, which was redesigned in summer 2021.

New buffer for painted vehicles.

Intelligent storage

The newly built paint shop and accompanying switch to a third shift made it necessary to have a high-bay warehouse for the painted vehicle bodies. It was the only way to ensure that the bodies painted in the night shift were buffered, and that the assembly division was supplied with the correct number of units. In addition, numerous conveyor bridges, more than half a kilometre in length, were built as a supply route between the new high-bay warehouse and the assembly division.

Only one site offered sufficient floor space for a storage facility with around 800 bays. This was where the high-bay warehouse was to be built. However, the location necessitated a rather untypical construction method for such a warehouse. The entire complex stands on stilts and therefore on an around eleven-metre-high platform. The main logistics route runs underneath it.

The high-bay warehouse (without superstructures) is the tallest building in the factory with a total height of 41 metres. Its special feature is the visible concrete façade, giving it a rough appearance, which is unique in the factory.

800 storage spaces are located in the 41-metre high-bay warehouse.

The plant management and planners are in on the action

While the existing assembly buildings were being expanded, between 2016 and 2018 another entirely new building was also constructed at the heart of the plant. In addition to the extra space for assembly and logistics, it includes new offices on the two upper floors, which offer the plant management and various planning departments ideal conditions and modern working environments. Building 16.3 was integrated as a three-sided extension onto a previously created open space. The passages between the buildings were opened gradually in order to create the impression of a consistent blend between the new and old buildings.

The start of vehicle assembly is located in the new central building 16.3.

View of the four-cylinder from the new office space.

The newly created production areas will be used, among other things, to install components for the 48-volt on-board electrical system. In addition, space has already been made available to integrate vehicle models such as the all-electric BMW i4 into the production line.
The newly created office space provides the ideal conditions for office and production staff to connect. The plant management's open integrated workspaces send a positive message. The managers are visible and approachable for all employees. Hierarchical boundaries are broken down and a new, agile form of collaboration is promoted and experienced.
The offices are well designed, with attractive furnishings and balanced materials and colours making the corporate values tangible while having a positive effect on the working atmosphere. There are no assigned workstations, but the "flexible office" model means the workspaces can be used in the best possible way.

Space for exchange and networking in the new central building.

BMW 316
BAYR. MOTOREN WERKE

Built to shape tomorrow: the BMW Group's Munich plant cele- brates its centenary

The home plant in Munich, with its long history, represents the company's success and transformation like no other BMW Group site can. The challenges at the company's only city-centre production site are enormous: increased urban densification and more difficult logistics and supply conditions. The main plant is being adapted for "New Class" production, just in time for its 100th anniversary. This transformation is based on three focal points: transformation in terms of e-mobility and digitalisation, increased efficiency of processes and structures, and enhanced sustainability in production and logistics.

2020 –

A promising decision

Shortly before the turn of the year 2020/2021, the BMW Group announced the decision to gradually transfer its long-established powertrain operations from the Munich plant to the Steyr (Austria) and Hams Hall (UK) sites.
This is a necessary step to secure vehicle production capacity at the BMW Group's head office. New and additional production structures are needed as the models increasingly turn electric, something which can no longer be accommodated in the current buildings and workshops. Therefore, engine building gave way to new state-of-the-art assembly lines entirely dedicated to electric vehicles.
The roughly 1,200 employees of the powertrain division were personally told about the changes by their managers and in a joint event by the Board of Directors, plant management and works council. A comprehensive transformation programme was immediately put into effect, as every employee is to be given another job at the Munich plant or at another BMW Group site by the time the last engine assembly line is finally packed up at the end of 2023.
The demolition of the building followed immediately after the powertrain operations were moved. While the demolition of the western part was still underway, construction on the new building began at the opposite end. The ambitious plan is for the first vehicles to start being commissioned and produced just over two years after the first demolition day.
At the same time, a new body shop is being built on the site of buildings 19 and 20, which were decommissioned in 2018. Hall 19 was built in the mid-1920s and had a variety of functions over the years: first as an assembly hall for motorcycle production and most recently as a paint shop. The two new buildings will enable series production of the "New Class" at the Munich plant from 2026. Their vehicle architecture is completely new and is uncompromisingly optimised for electric drives and follows three key aspects: redefined IT and software architecture, newly developed electric drive and battery generations, and radically new sustainability levels across the whole vehicle life cycle.

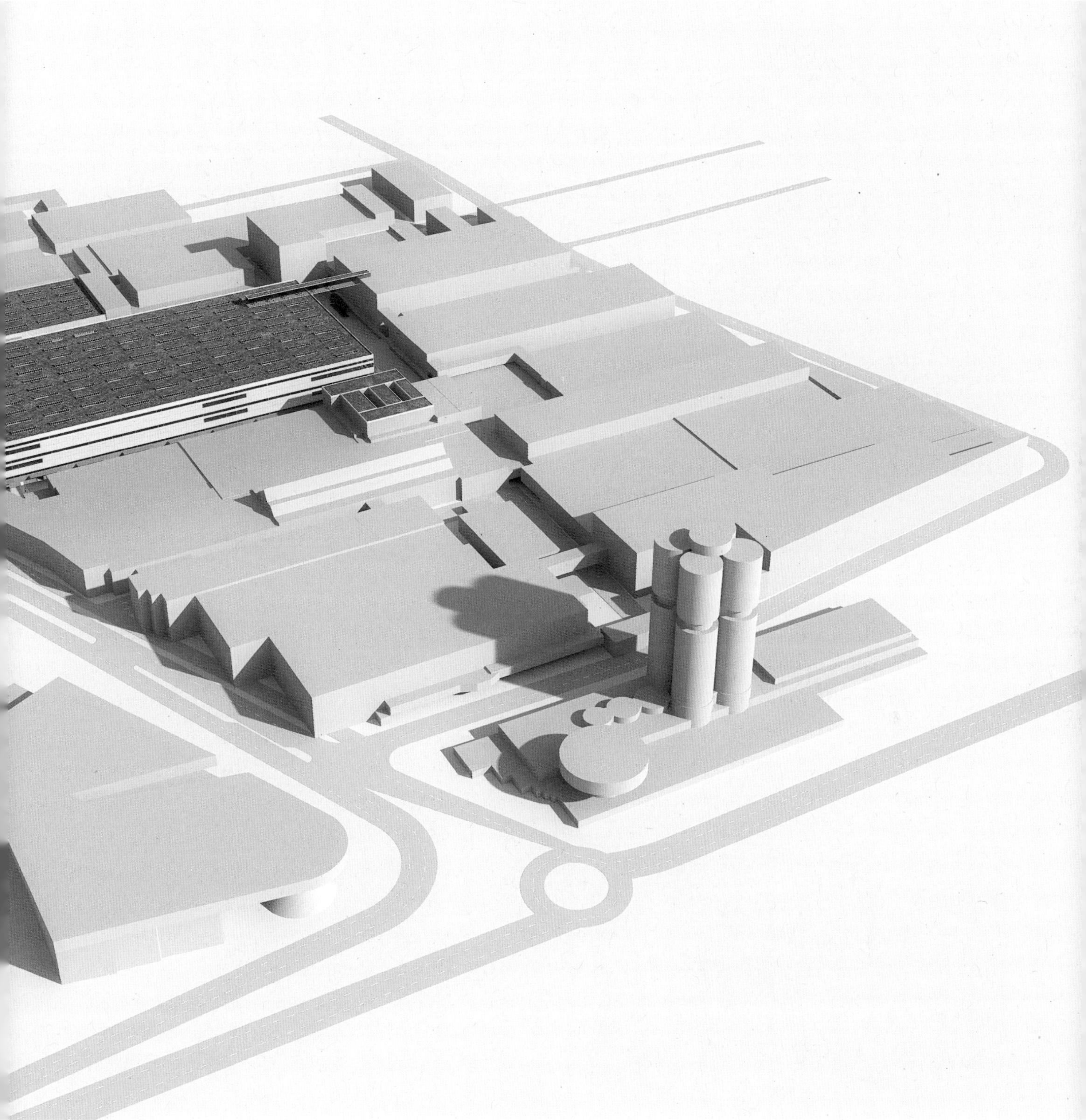

A new state-of-the-art assembly will be built on the area of the engine plant by 2026.

Mentor Gashi

I'm a trained specialist in Automotive Mechatronics and have worked in engine-building at the Munich plant for years now. At the end of 2019 our management told us how powertrain operations will develop over the next few years and what opportunities we have. I thought hard about my future, and ultimately applied for the body shop based on a pure gut feeling.
I used to think it was loud and sweaty in the body shop, that you worked with huge equipment and machines – but it's very different in the body shop finishing plant. It's so multifaceted, you learn about the finishing touches, the details – there's so much behind it! I have to concentrate more here, it's more demanding, because if something doesn't work, you have to come up with a solution. I like that. The trial work was actually planned to be a month long. But I already knew I wanted to stay after two weeks. Right from the start, they made me feel like I was being heard and that it mattered how I was doing. It was the right decision to switch, and I've also told my former colleagues in engine building that.

As a farewell to the historical halls, two vehicles returned to their roots: the R32 Motorbike, which was produced in this hall in the mid-1920s, and a BMW 3 Series of the sixth generation, both of which were painted in Hall 19 until 2018. Plant Manager Robert Engelhorn, Manfred Schoch (Chairman of the Joint Works Council), Nicole Haft-Zboril (Director of Real Estate Management) and Wolfgang Obermaier (Director of Indirect Goods Purchasing and Services) (from right to left) ushered in the demolition in June 2021.

Architecture competition

Major structural changes to the plant resulting from the two large-scale assembly and body shop projects provide an opportunity for further architectural developments. The plant's unique urban location has to be accounted for. So, the entire complex is to be made noticeably more accessible and open to the neighbourhood.
In October 2021, the BMW Group launched the international architecture competition "BMW Munich – Urban Production" for the long-term structural development of the home plant in Munich. The essential project goal is to transform the plant into an urban-feeling production campus with international appeal.

Search for ideas for the Munich plant as a production campus in an urban environment.

Surrounded by residential areas and the Olympic Park: A goal of the architectural competition was to create the opening of the factory to the neighbourhood.

In the east, west and north, residential areas border the plant.

Six renowned international architecture bureaus took part in the competition, and the results were announced in February 2022. The two winners OMA (Rotterdam) and 3XN (Copenhagen) won over the high-calibre jury, which included the Lord Mayor of Munich, Dieter Reiter, and the BMW Management Board. The entire process was carried out in close cooperation with the City of Munich. To keep the neighbourhood updated about the project, events were held with residents living around the plant and their suggestions and wishes were taken on board.
Plant manager Peter Weber said on the launch day: "In order to remain viable in the long term, overall change is necessary. The first two major projects are the new vehicle assembly and a new body shop. We're therefore creating the basis for innovative, sustainable and competitive production processes. We are preparing our plant to produce the "New Class" models, which is our next step into the third phase of e-mobility. With the architecture competition, we want to create a way of potentially opening up the plant in Munich's city centre. The transformation of our plant will help secure the future of the site and its jobs."
BMW's Director of Real Estate Management, Nicole Haft-Zboril, added: "We want to fulfil our social responsibility with this project, and actively participate in developing the neighbourhood surrounding the plant. This will create a lasting future-oriented work and production environment. In this way, we are taking an integrative approach, especially with regard to transport links, the neighbourhood and sustainability."

GEBAUT, UM DAS
MORGEN ZU GESTALTEN.

100 Jahre Werk München.

Werk 1.1 Tor 1

BMW GROUP

Campaign theme for the activities in the anniversary year of the Munich plant.

One look into the past, two into the future

The timing for redesigning the plant's premises could not have been chosen better. In 2022, the Munich plant will celebrate the centenary of its creation. The previous chapters of this book show clearly that the last one hundred years have been very eventful, with many unpredictable twists and turns, dramatic decisions, anxious moments, but also great successes. In 2022, the Munich plant is in a much better position than many in the past would have thought possible.

The workers at the Munich plant can feel on a daily basis how history and future of production go hand in hand – the plant's workplace offers a historic feel combined with state-of-the-art production facilities. This unique feature has to be preserved as well as developed for the future. That's why "Built to shape tomorrow" is the slogan for the centenary year 2022.

The beginning of a new era

A new era is being ushered in at the Munich plant shortly before the start of the centenary. The BMW i4 is the first all-electric Gran Coupé to roll off the home plant's production lines. Leading up to the start of production in the autumn of 2021, the plant and its workforce were prepared for e-mobility via the largest assembly conversion in history, plus an extensive qualification program (see chapter 11, page 258).

Preparing for the start of production of these new models was already an intense time, yet this was further aggravated by the COVID-19 pandemic. The Munich plant was also shut down during the nationwide lockdown in spring 2020 and the production downtime planned for Easter was brought forward and extended – for eight weeks, not a single vehicle was produced.

When production finally resumed on 18 May, initially operating with a single shift, numerous measures were introduced to protect the staff from infection at work. During the production downtime, the company manufactured partition walls in-house for work and break areas, installed social distancing markers, set up hand-sanitiser and mask dispensers, drew up occupancy plans, and much more. There was a great deal of discipline among the workforce; everyone wanted to start working and producing vehicles again. In the subsequent COVID-19 waves up to 2022, the measures proved their worth, despite being relaxed in the meantime and then tightened again, and were widely accepted, meaning

Dr. Dorothee Lange-Rieß, Head GP at the BMW Group's Munich plant

On 11 March 2020, the World Health Organization classified the spread of the SARS-CoV-2 virus as a pandemic. We didn't have any experience of pandemics despite all the existing medical expertise. But the virus had already been a fixture in the BMW health services' day-to-day activities from January 2020, at first mostly in relation to medical travel advice prior to business trips. When the COVID-19 pandemic threatened, the focus then quickly shifted to occupational health and safety for employees and ensuring the company's ability to continue working.
To support the health authorities, our health services helped with case and contact tracing of employees who had fallen ill, in order to quickly identify and interrupt chains of infection. We not only held many medical consultations over the phone, but also clarified questions and worries with those affected.
High standards of health protection were consistently implemented and kept track of in the company. This was only possible by working closely with all process partners and through intensive interdisciplinary and cross-site networking – whether in the initial "COVID inspections" or by establishing (local) expert teams who constantly adapted and implemented measures against infection. They, of course, always observed the constantly changing official and legal requirements in their local regions.
Back in December 2020, we had already launched an interdisciplinary project team to prepare for the health services' COVID vaccination campaign. Within four months, vaccination centres complete with GP consultation were up and running at BMW plant sites and were designed to avoid unnecessary disruption to production. It was wonderful to see how gratefully our colleagues received our comprehensive vaccination services.
All in all, we were always in control, despite the challenges. This was due to high levels of commitment, motivation and flexibility, and also thanks to our intensive in-house cooperation.

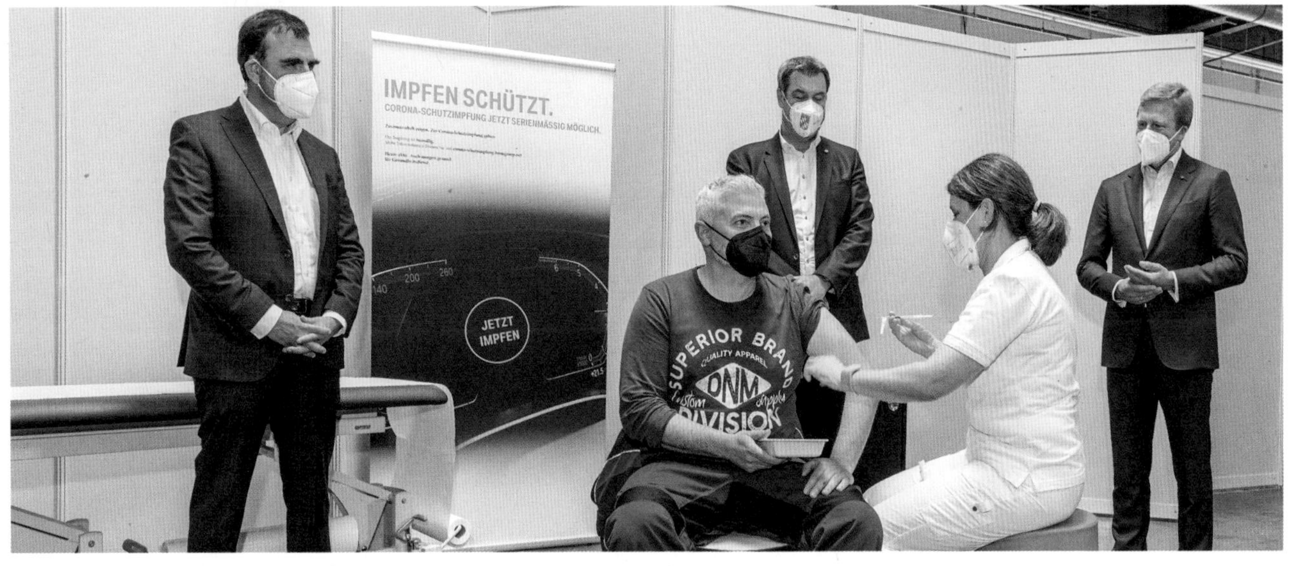

Bavaria's Prime Minister Markus Söder attends the opening of the vaccination centre at the Munich plant.

that the Munich plant saw no major outbreaks during the entire course of the pandemic.

The BMW Group's extensive vaccination programme was a great help in this. From June 2021 there were a total of 29 vaccination centres running on German BMW sites, with 14 of them at the Munich plant and in the FIZ (Research and Innovation Center). After the first and second dose rushes in the summer, in the autumn the vaccination centres were given over to the general health services, where booster jabs are still currently being offered. At the peak, up to 2,500 vaccinations a day were being given across the whole of BMW.

Production Board member Milan Nedeljković, plant manager Peter Weber and works councillor Elisabeth Altmann-Rackl with the first BMW i4 on the day of production start.

The efforts made during this eventful time leading up to the start of production paid off: the BMW i4 was started ahead of schedule and was a success story from the beginning. The press and social media were enthusiastic about its driving dynamics and efficiency; the orders from dealers and customers exceeded the already very positive expectations. Production was therefore quickly ramped up and the traditional downtime over Christmas was even shortened in order to make up for shifts that had initially been cancelled due to the semiconductor shortage, and to supply customers as soon as possible.

The BMW i4

With the BMW i4's go-ahead, pure e-mobility has reached the heart of BMW's brand. The four-door Gran Coupé combines the spacious comfort and practicality of this vehicle concept with the on-brand sporting flair and an impressive mileage even for long-distance journeys. The BMW i4 is the brand's first purely electric model, with a consistent focus on driving dynamics from the outset of the design. The BMW i4 eDrive40 combines a 250 kW/340 hp electric motor with classic rear-wheel drive. Its mileage, determined in the WLTP test cycle, is up to 590 kilometres.

The i4 M50 model variant is also the first BMW M vehicle with a localised emission-free drive system. With an electric motor on each of the front and rear axles and a system output of 400 kW/544 HP, BMW M GmbH's performance car generates intense joy while driving with a range of up to 521 kilometres according to the WLTP test cycle.

Water power

The car itself is not the only example of the BMW Group's uncompromising sustainability strategy. The production of the BMW i4 is also steering the plant to make further sustainable changes. Since 2020, the BMW Group everywhere has only been using green electricity. To produce the BMW i4, the company has signed supply contracts for renewable electricity with the power plants of Gersthofen and Rain in the Lech region, to further increase the environmental quality of the green electricity used for production.

Water is also key to a new system in the Munich plant's paint shop: with the help of a reverse osmosis system, fresh water consumption has been significantly reduced. Here, wastewater from cathodic dip painting – a type of primer for vehicles – is also treated and is then reused in the same processing step. Overall, this reduces annual fresh water consumption by more than six million litres. Since 1997, the plant has been obtaining around half its annual water volume for production processes from its own on-site well, therefore conserving valuable drinking water.

The Gersthofen hydroelectric power station on the Lech Canal was put into operation as early as 1901. Now it supplies green electricity for the production of the BMW i4.

Each vehicle produced by the company has already had its resource consumption more than halved in the period from 2006 to 2020. CO_2 emissions have fallen by 78 percent. By 2030, CO_2 emissions per vehicle produced are set to fall by a further 80 percent.

A new sustainability goal: emission-free local transport connections

Another step towards increasing sustainable production is emission-free logistics of parts and vehicles in the local area. More than 750 truck deliveries are required every day to supply parts. In the future, these journeys in and around urban areas are to be made emission-free using e-trucks. About half the vehicles produced in Munich are already transported by rail, and this proportion is to be gradually increased. Effectively, the plan is to implement zero-emission transport logistics in Munich's urban area and to keep reducing CO_2 emissions significantly in the wider area and with intercontinental traffic.

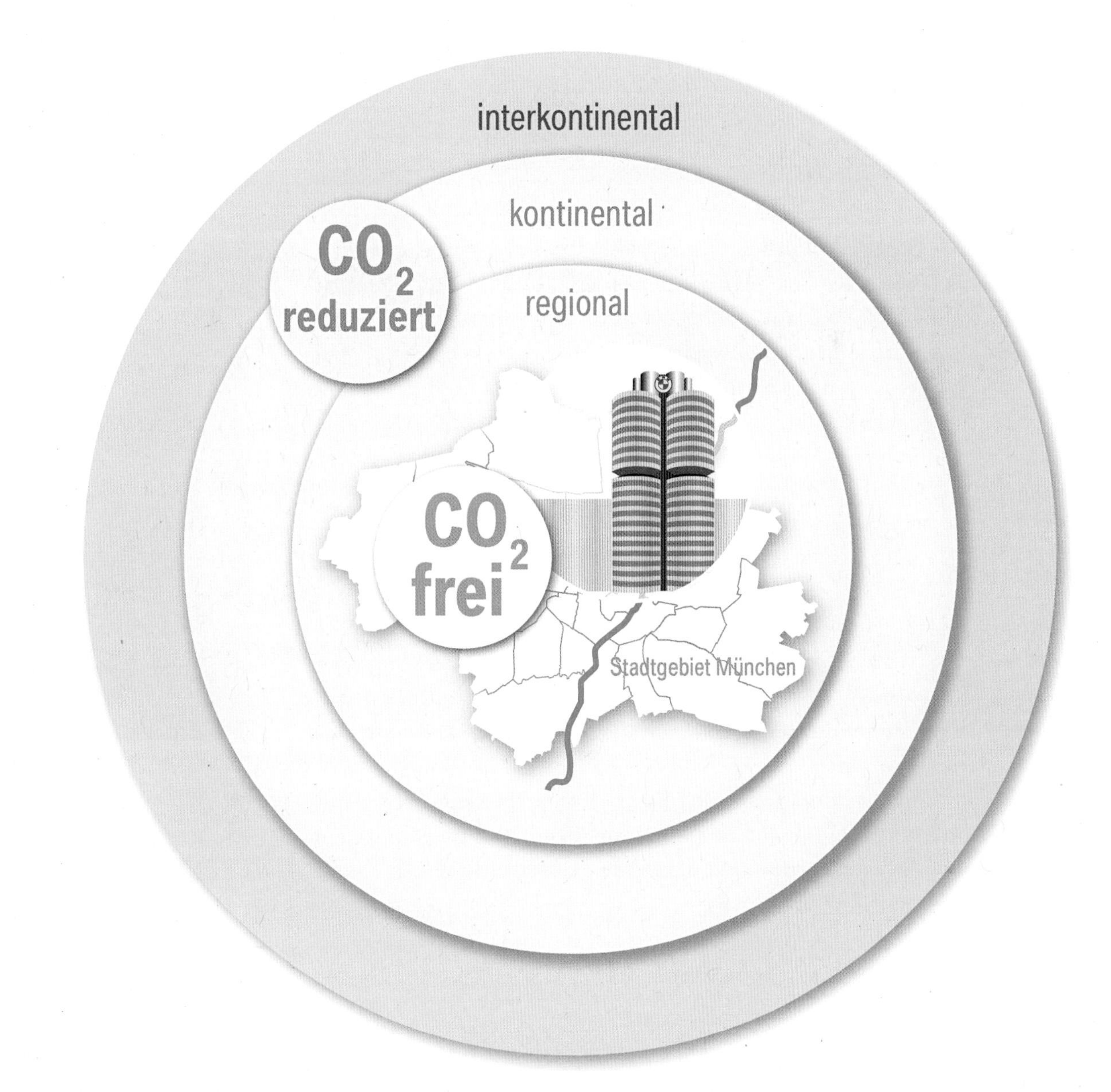

Still central

The home plant is still firmly anchored within the constellation comprising the "Four-Cylinders" tower which serves as corporate headquarters, visible from afar, the BMW Welt, the BMW Museum, the Research and Innovation Center, and BMW Group Classic, which was newly occupied in 2016. The six locations shape the cityscape of Munich's north with their extraordinary architecture – each in its own right and as a collective window into the company's more than one-hundred-year history.

The company headquarters

"The biggest four-cylinder in the world" stands as a symbol for the city of Munich and in 2013 was accepted, by a commission of experts, into the ranks of the fifteen most spectacular corporate headquarters in the world. Designed by architect Professor Karl Schwanzer, the tower was officially inaugurated on 18 May 1973, and has since become an undeniable icon of architectural history: short distances on the inside and clean lines on the outside. The corporate headquarters and museum have been registered as protected sites of historic interest since 1999.

Landmark of the city: the BMW four-cylinder, the BMW Welt and the Munich plant.

Flexible and transparent: The concept of COOP HIMMELB(L)AU architects is characterised by the hitherto unique roof construction and the so-called double cone.

BMW Welt

The plant's most recent neighbour is BMW Welt, where visitors from all four corners of the world have been able to learn about the BMW, Mini, Rolls-Royce and BMW Motorrad brands since 2007. BMW Welt realises the idea of creating space for encountering the new and has the task of bringing information to life for customers and visitors alike.
At BMW Welt, customers can collect their new BMW in a unique way and, with all five senses, experience what makes the BMW Group brands so unique. BMW Welt is now the most visited attraction in Bavaria – and at a site which means so much to BMW.

Think tank and nerve centre: The BMW Group Research and Innovation Centre "FIZ".

The Research and Innovation Center

The Research and Innovation Center (FIZ) currently has the most employees out of all the BMW premises. Started in the mid-1980s and inaugurated in 1990, the building complex has continued to grow and is home to engine developers, designers, car body specialists, chassis experts and electronics specialists – as well as physicists, chemists, materials researchers and software engineers. Additionally, buyers, production planners and construction specialists work on the site, which is modelled after a university campus, along with many other specialists required for the development process of a modern vehicle. Every model for every BMW Group brand and market is conceived at the FIZ. It is the company's think tank

and nerve centre, where all the important information from global research and development networks is amassed and decisions are made around new technology and vehicle concepts.
In recent years, the architecturally impressive building complex has become one of the largest R&D locations in Europe due to its continuous expansion over the course of the urban development project "FIZ Future".

Architectural design for "FIZ Future".

The characteristic "museum bowl" was designed by the architect Karl Schwanzer.

The BMW Museum

The BMW Museum in Munich brings over 100 years of automotive interest, innovation and driving pleasure to life and answers questions about BMW's corporate, brand and product history in its exhibitions. These span from the beginning to the present and into the future, giving museum guests a comprehensive picture of the BMW brand's innovative strength.

And it works too: with 630,000 visitors per year, the BMW Museum is one of Munich's most popular museums. The building is visually impressive due to its unusual architecture. The characteristic "museum bowl" was designed by the architect Professor Karl Schwanzer to "continue a street-feel in an enclosed space". In 2008, the museum was redesigned and is now five times larger, comprising 5,000 square metres of exhibition space.

Protected monument: The gate building of the BMW Group Classic.

The BMW Group Classic

Since 2016, Moosacher Straße 66 in Munich has been the first port of call when it comes to classic BMW Group vehicles. On the 13,000 square-metre site, experts take care of all matters relating to the historic vehicles, such as restoration, replacing parts and archiving information. The BMW Group Classic museum also exhibits some of the now 14,000-strong collection of vehicles. Here another full circle is completed, as the collection of historic buildings also includes one of the first production halls on the site of the then still young Munich BMW plant. It was preserved during its remodelling work, as was the contemporary gate building, which now has protected status. This building now serves as a gateway to the BMW Group's history, thus making the brand's past once again part of its future.

One of the first production halls of the Munich plant was on the premises of the BMW Group Classic, which opened in 2016.

Publication details

Published by	BMW Group
Authors	Caroline Schulenburg (Chapters 1–7) Andreas Hemmerle (Chapters 8 -11) Susanne Tsitsinias (Chapters 11 and 12)
Concept and realisation	von Quadt & Company – Agentur für Kommunikation und Events, Hirmer Verlag (2nd and 3rd editions; new design for the 3rd edition: Lucia Ott)
Photography	BMW Group Historical Media Archive BMW Group PressClub BMW Group Munich plant Siemens Archive Pages 154, 168, and 186 © Werner Bachmeier Page 284 LEW / Yeah - Bild, Code & Herzklopfen GbR Marcus Buck, Peter Rossa, Andreas Hemmerle, Harry Zdera, Günter Schmied
Translation	Sonia Brough, Michael Capone, Tristam Carrington-Windo, Hugh Keith, Timothy Kemp, Philip Radcliffe, Alan Seaton, Micha Goebig-Phelps
Lithography	Reproline mediateam GmbH & Co. KG, Munich
Printing and processing	Westermann Druck, Zwickau
Paper	120 g/m² enviro clever
ISBN	978-3-7774-4073-6 (Hardcover) 978-3-7774-4076-7 (Softcover)

3rd revised and enlarged edition 2022